DATE DUE

DE 18 98			
NO 2 99			
NO 20 02			
DE 5 02			

DEMCO 38-296

BEYOND EVOLUTION

Beyond Evolution

Human Nature and the Limits of
Evolutionary Explanation

ANTHONY O'HEAR

CLARENDON PRESS · OXFORD

1997

at Clarendon Street, Oxford OX2 6DP
d New York
Bangkok Bogota Bombay
Buenos Aires Calcutta Cape Town Dar es Salaam
Delhi Florence Hong Kong Istanbul Karachi
Kuala Lumpur Madras Madrid Melbourne
Mexico City Nairobi Paris Singapore
Taipei Tokyo Toronto Warsaw
and associated companies in
Berlin Ibadan

Oxford is a trade mark of Oxford University Press

Published in the United States
by Oxford University Press Inc., New York

British Library Cataloguing in Publication Data
Data available

Library of Congress Cataloging in Publication Data
Data available
ISBN 0–19–824254–9

1 3 5 7 9 10 8 6 4 2

Typeset by Hope Services (Abingdon) Ltd.
Printed in Great Britain
on acid-free paper by
Biddles Ltd.,
Guildford & King's Lynn

For Thea

Preface

THE aim of this book is to examine the extent to which evolutionary accounts of human experience are adequate. In examining this question, I focus on human knowledge, on morality, and on our sense of beauty. I suggest that our current activities in each area certainly derive in important ways from our biological nature, but that once having emerged they cannot usefully be analysed in biological or evolutionary terms. I also attempt to indicate the significance of human community and of our cultural inheritance in the identity and rationality of each one of us. At the same time, I attempt to vindicate the traditional view that each human being is possessed of a rationality which means that he or she can transcend what is given in biology and culture. We are prisoners neither of our genes nor of the ideas we encounter as we each make our personal and individual way through life.

Early versions of parts of the book have been previously published as follows: part of Chapter 2, as 'Immanent and Transcendent Dimensions of Reason', in *Ratio*, 4 (1991), 108–23; part of Chapter 3 as 'Self-Conscious Belief', in *The Certainty of Doubt*, edited by Miles Fairburn and Bill Oliver (Victoria University Press, 1996), 336–51; part of Chapter 4 as 'Evolution, Knowledge and Self-Consciousness', in *Inquiry*, 32 (1989), 127–50 (reprinted by permission of Scandinavian University Press, Oslo); part of Chapter 5 as 'Knowledge in Evolutionary Context', in *International Studies in the Philosophy of Science*, 8/2 (1994), 125–38; part of Chapter 7 as 'Beauty, Natural and Unnatural', in *Artists from Europe*, edited by Kevin O'Brien (Leeds Metropolitan University, 1995), 72–8; part of Chapter 8 as 'Two Cultures Revisited', in *Verstehen and Humane Understanding*, edited by A. O'Hear (Cambridge University Press, 1997), 1–21. Where appropriate, I thank editors and publishers for permission to reprint.

As will be evident, I have worked on this book and related themes over a number of years, and have benefited from many discussions with many people for ten years or more. It would be hard to list all who have helped me in one way or another, though I would like to mention my gratitude for particular help I have received from Roger Trigg, Michael Smithurst, Peter Munz, and Roger Scruton.

Contents

1

Mind and Nature

In considering the interaction between human beings and their material surroundings, we cannot help but be struck by two things. First, and most obvious, human beings are themselves material beings, acting in and on, and produced and influenced by the material world. In saying this, I do not mean to pre-empt discussion of or conclusions about minds, souls, or spirituality. I simply want to make the obvious point that we are embodied, and that what we do and think and feel is conditioned by our embodiment. In our knowledge, in our morality, and in our perceptions of beauty, our materiality is presupposed and exploited in a host of ways, some of which I hope to explore in the course of this book. But even before we go any further, we can take for granted that our senses condition and filter our knowledge, in interaction with the material world; that morality has a lot to do with human suffering and pain due to physical limitations; and that aesthetic experience is intimately linked to the perception of visible, aural, and tactile things, and also to phenomena smelt and tasted.

So any consideration of human nature will have to involve an examination of our materiality, both as it is and as it has come about, presumably in evolution. But, and this is the second point which immediately confronts us, it is also obvious that in various respects we do not behave like most material objects. We are conscious, which most of matter is not, although consciousness is something we share with a good proportion of the animal kingdom. Over and above consciousness, though, human beings have developed the ability to think discursively, to be aware that they are so thinking, and to express these thoughts in language and other symbolic forms. As the concrete expression and development of this propensity for self-conscious thought, we have produced all kinds of cultural artefacts and systems, by which our lives are surrounded, conditioned, and made meaningful. No doubt, as we shall see in greater detail in due course, human culture and the world in which we live—the *Lebenswelt* of the phenomenologists—draws on continuities between us and the lower animals. No doubt, too, our comparatively large and complex brains play a crucial role in the production of thought and culture, and in everything which follows we shall assume that this is so. Nevertheless, human thought and culture do exhibit certain important properties, rare if not unique in the rest of nature. Saying this does not in itself mean that they cannot be given a naturalistic, or, more precisely, an evolutionary explanation. The reason for focusing here on evolutionary explanations is that evolution is the generally accepted account of how species in the biological kingdom have developed the physical structures and behavioural dispositions they have. If we see a species as part of nature, we will, then, be committed to giving an evolutionary account of its origins and development. What this book is about is exploring the strengths and weaknesses of evolutionary accounts of human activity, which any naturalistic explanation of human activity will have to face, but which some recent and sophisticated accounts of human activity tend to obscure.

But we cannot begin to make progress here without getting clearer about the nature of our abilities and about the relevance to them of evolutionary explanations. There is no better or more appropriate place to begin this task of clarification than Socrates' discussion of the relationship between the mind and nature in Plato's *Phaedo* (95–101). For although Plato does not treat of evolutionary explanations specifically, what he does is to present a hurdle any naturalistic explanation will have to cross.

Phaedo

The dialogue, it will be remembered, takes place on Socrates' last day on earth. It ends with Socrates' death, and with his famous last words: 'we ought to offer a cock to Asclepius'. Asclepius is the Egyptian god of healing, whose cult had latterly been taken up in fifth-century Athens. So what the words imply is that life is a disease and death the cure, releasing us into a healthier and more enlightened form of existence. In *The Birth of Tragedy* Nietzsche had found these words and their implication particularly distasteful, a symptom of a life-denying, quasi-religious attitude to this life. But while Platonic dualism and Platonic pessimism are certainly in the background, the passage we are about to consider (95–101) does not presuppose Platonic metaphysics. What it does presuppose, and what certainly might raise questions for a Darwinian, is the thought that a certain kind of life might be so shameful that death is preferable.

In our inquiry in the course of the book we will frequently be confronted by perplexities of the sort just mentioned, that some aspect of our life or thought poses a question for a certain type of explanation of our existence. In this case, the tension is between the feeling of shamefulness and the evolutionary drives to survival and reproduction. Of course, the Darwinian is not without recourse here. He can give an account of the nature of shame in his own terms, showing it to be an adjunct of pride and self-respect, attributes which may well, in the main, contribute to an individual's ability to survive and reproduce. This, though, will not show why, in the specific case Socrates felt that he had to die rather than survive; and if it is said that from the evolutionary point of view Socrates was a failure, and that a nation of Socrateses would not survive (leaving aside the fact that he had already produced children), this still does little to explain why Socrates is so widely admired a figure, and why many people even today, and in quite different social and religious circumstances, feel that Socrates was right to have done what he had done. Here we have the germs of a conflict, between a doctrine which says that there are considerations which ought to override survival and reproduction, and one which says that ultimately rational morality must have its edicts legitimated in terms of their contribution to the survival and reproduction of individual agents, actual and potential.

It might be said that strictly a scientific theory, such as the theory of evolution, does not contain within itself the resources to underpin or, indeed, to undermine a normative activity such as morality. This is doubtless true, so long as we are looking for the theory to tell us directly what we ought to do, but it would also be too quick a way of completing the present exercise. For, as an explanation of human behaviour, the theory of evolution should not be read as telling us directly what we ought to do, but rather as explaining why we engage in the normative activities in which we do engage, and explaining those norms in terms of their contribution to survival and reproduction. Here the possibility of conflict does arise, if we discover that humanity, or large and successful tracts of humanity and what are widely regarded as the best of humanity do have norms and practices which are both significant in their lives and thought and either indifferent to survival in their effects, or actually counter to survival. If we are Darwinians about human behaviour we will want—as Darwin did in *The Descent of Man*—to know how far the apparently survival-indifferent practices are only apparently survival-indifferent; failing which, we will have to concede that Darwinism is incomplete as an explanation of human behaviour, and that there are aspects of human behaviour which do not just need supplementation from other data for their explanation, but which point to non-Darwinian sources of reasoning and motivation within us. For it will not simply be that there are drives in us which are irrelevant to survival, or, as I will suggest, actually counter to survival. As rational beings, we are not simply the victims or creatures of drives. We act, at least in part, for reasons; and if the reasons are un-Darwinian, unselfish, unconcerned with individual survival and reproduction, we will want to know why *those* reasons carry weight with us, what their sources are, so to speak, why they seem to be good reasons, and ultimately whether they actually are good reasons.

If it is the case that we have non-Darwinian motivations, as I am suggesting the example of Socrates' last discourse might show, then it is, of course, still open to the Darwinian to say that we had better revise our reasoning and amend our motivation, or destruction rather than survival and reproduction (or survival through reproduction) faces us. But, while as a matter of fact this may be true, and destruction ensues, given that we do have Socratic and other non-Darwinian reasoning and motivations, we cannot without begging the question in favour of a Darwinian imperative simply accept that this is what *should* be done. We would now have a conflict of reasons and a conflict of oughts, and it remains an open question as to which we should choose without doing ourselves the greater damage, or indeed, as to what might count as doing ourselves damage.

This, though, is to rush ahead. We need first to examine Socrates' speech and our everyday morality, in so far as that follows Socrates so as to see just what it might be about them which conflicts with a straightforward naturalism like Darwinism. (In my analysis and interpretation of the speech, I am indebted to Stanley Rosen's brilliant discussion of it in *The Question of Being*,[1] though my purposes and conclusions are quite different from his.)

Overall, the speech tells of Socrates' turn from natural science to philosophy, and of his realization that there are different kinds and levels of explanation. What started as an 'extraordinary passion' to discover the causes of things, and what took Socrates into the natural science of his day quickly led him out of it and into philosophy,

---

[1] Stanley Rosen, *The Question of Being* (Yale University Press, 1993), 46–95.

convinced that he was 'uniquely unfitted' for science. Socrates' initial passion was to discover causes, and whether, for example, living creatures arise from fermentation, or whether thinking is caused by blood or air or fire or the brain. But he finds himself quickly becoming bogged down in the contradictory and paradoxical nature of the speculations involved. Now, one could certainly reply to Socrates at this point that this in itself is no reason for despair. Scientific speculation has often proceeded by means of resolving paradoxes and contradictions, perceived or real. Perhaps especially today, where we are beset by the paradoxes of quantum theory and the conundrums of astrophysics, much of the motivation for scientific activity is the need to remove unclarity and difficulty from existing theories. This does not necessarily involve making what is being explained more familiar or easier for untutored common sense. Precisely the opposite may well be the case: scientific advance, as Popper has remarked, is often in the opposite direction, from the known to the unknown, revealing the world as stranger than ever thought.

Part of Socrates' objection to scientific speculation focuses on just this point, the way in which science moves from what is clear and familiar to what is obscure and unfamiliar. What Socrates will argue is that in the sort of case he is interested in, he will do better to rely on what he knows from the start. The truth here lies not, as with science, in strange and obscure speculation, but in things he and we already know. One thing in particular which we know, and which scientific speculation only obscures, is that at times we have the intention to act for the best. In fact, it turns out that Socrates is not interested in scientific explanations at all, but only in the type of account which science, as normally understood, is specifically precluded from giving. What we have here is the basis for a distinction at a fundamental level between philosophy and science, and the basis for an argument against a view of philosophy widely prevalent at the end of the twentieth century.

That this is so becomes clear when, having spoken of science as a type of inquiry for which he was 'uniquely unfitted', and of his subsequent determination to 'muddle out a haphazard method' of his own, Socrates says that he once heard someone reading from a book by Anaxagoras, in which it was asserted that it is Mind which produces order and is the cause of everything. Now

this explanation pleased me. Somehow it seemed right that Mind should be the cause of everything . . . Mind in producing order sets everything in order and arranges each individual thing in the way that is best for it. Therefore if anyone wished to discover the reason why any given thing came or ceased or continued to be, then he must find out how much it was best for that thing to be, or act or be acted upon in any other way.

Anaxagoras seemed promising because it looked as if he was giving explanations in terms of how it would be best for something to be. In contrast to the muddling and befogging speculations which Socrates had turned away from, Anaxagoras seemed to promise a scientific account which would link both with our experience of human affairs and with our desire to look at things in the light of the purposes and intentions underlying them.

Anaxagoras thus seemed to put an end to the meaninglessness which otherwise bedevilled scientific and cosmological speculation; he seemed to promise an answer to the threat so frightening to the thinkers of the eighteenth and nineteenth centuries of an ultimate meaninglessness in a universe vast in extent and time, and utterly unresponsive to us and our sense of value. Unfortunately, however, far from fulfilling these hopes, Anaxagoras is revealed to be an archaic Stephen Hawking, one who simply describes the mechanisms of nature, and then tells you that in hearing all this you have read the mind of God (or in Anaxagoras' case, of Mind).

Socrates tells us that he assumed that a person inspired by the notion of Mind as the cause of everything would begin by describing such things as the shape of the earth or the velocities and orbits of the heavenly bodies, and then in some detail explain why it was best that things should be as they are: 'it never entered my head that a man who asserted that the orderliness of things is due to Mind would offer any other explanation for them than that it is best for them to be as they are.' But, on examination, Anaxagoras turned out to have anticipated by two millennia or so the contemporary scientific expurgation of teleological explanations. 'The fellow made no use of Mind, and assigned to it no causality for the order of the world, but adduced causes like air and ether and water and many other absurdities.'

Socrates goes on to draw a comparison between Anaxagoras' explanation of the world, and someone who tried to explain Socrates' own lying in prison, awaiting death, in terms of the actions of his sinews contracting and relaxing in such a way as to cause his bones and flesh to be where they currently are. Socrates says that this would be absurd. What causes him to be where he is is the thought he has that it is 'better and more right' to stay in Athens and submit to whatever penalty his country orders—otherwise his bones and sinews would long ago have taken themselves off to Megara or Boeotia. Indeed, to call the dispositions of his bones and sinews causes in the present context is also absurd. While, without his bones and sinews he would not be able to do what he thinks is right, what he does is through his choice of what is best; in like manner, he was looking in Anaxagorean cosmology for some reference to the world being directed and held together by something like goodness or moral obligation.

It is possible to read what Socrates says as a plea for teleological explanations, and this is certainly part of what is involved. Against that plea, it would be said that modern science removed teleological explanations from the physical sciences in the seventeenth century, and that this was part of the explanation for its success. In the biological and human sciences, teleology still has a place, given to it by what are known as functional explanations. Thus we can say that the heart is for pumping the blood, and that is part of the reason for its being there: the heart has, in evolution, been selected *for* this purpose. Similarly, it is argued that human truth-seeking and following of moral codes, and the concepts involved in these activities have also been selected for, as promoting our survival in various ways. We will look at explanations of this sort in more detail later, but even at this point it is clear that these are not the types of explanation Socrates is interested in. The heart may be for pumping the blood, and pumping the blood may conduce to our survival. But is it good that we should survive? Equally, morality and scientific inquiry may conduce to various types of human life and activity; but are these activities themselves for the best? And is scientific research a worthwhile end if it does not fulfil any utilitarian functions or (in this ecologically conscious age) even if it does?

Socrates may seem to be looking for teleological explanations, and, in a sense he is. In thinking of physical events as controlled by Mind, he is certainly expecting that the preferred account will explain the whys as well as the wherefores of the events in question. But he is not interested merely in a teleology which does not directly lead on to moral questions. What he wants is an account which shows what it is that is for the best and why it is for the best, in the same way that he can show that it is for the best that he languishes in the condemned cell in Athens, rather than allowing his bones and sinews to take themselves off to Megara or Boeotia. Crucially, of course, he would not be in his cell were it not for the best that he is there. He could have escaped, but did not. If, as evolutionary theory suggests, we are programmed to be interested mainly in survival and reproduction, why did he not escape? Socrates would say that it was because he saw that it was for the best, something that evolution is going to have to offer a rather forced account of. Can an explanation in terms of what is for the best also be given in respect of the way the world is?

Modern opinion would be almost unanimous in saying that it cannot. To think that it could would appear to commit one to showing that this is the best of all possible worlds, or at least that it was such as to be directed by a good and all-powerful Mind. It would seem to entail that we show that the world is such that, for all its apparent imperfections, it 'really is bound and held together by goodness and moral obligation'. To the modern mind, the prospects for theodicy are not good; and even those who tentatively offer theodicies tend to do so in a defensive spirit, showing negatively that the world is not formally incompatible with moral goodness, rather than that it is positively directed by it.

Nevertheless, it is clear that for Socrates philosophy is fundamentally a moral activity, at least in an extended sense of moral. That is to say, it is to focus on how we ought to live and how we should investigate and evaluate things; it should focus on filling out the meaning of the best and of acting for the best.

It is at this point that Socrates makes more contact with tendencies and controversies in modern philosophy. There is a view of philosophy, associated with Quine and his followers, according to which philosophy should be seen as continuous with natural science, and a development of it. On this view, the traditional philosophical concerns, of providing justifications and foundations for our various activities and inquiries have been found unavailing partly because we have no way of stepping outside our activities. We must recognize that we have no Archimedean point from which to survey what we say and do, and see how it matches or fails to match the world in itself. If philosophy is to renounce any attempt to provide foundations, what it can do is to work out the implications and presuppositions of the scientific world-view.

Thus, epistemology ceases to be the attempt to reduce what we believe self-evident to truths of reason or to unquestionable, uninterrupted sense impressions and so, in a sense, to provide a criterion of truth for our beliefs. Instead it becomes a non-judgemental elucidation of the way in which the world, as given in natural science, actually produces our sense impressions and beliefs, and how far our beliefs and sensory dispositions are conducive to our survival or to our systematization of experience or both. In the latter activity, it is assumed that there is an external world, and that it is much as natural science describes it. The work of philosophy in general—or of naturalized epistemology in particular—is to explain the way the world *via* our senses gives us the representation of it we have. Part of the strategy here will be to show that we have the types of representation we do because they conduce to our survival, and so have been selected by natural selection, which favours and reinforces survival-promoting traits. This will, then, be some account of *why* our senses are by and large reliable, but it will be one which—by taking as its basis the theory of evolution together with neurophysiology and psychology— assumes that large tracts of our knowledge claims are in fact reliable. From the point of view of Socrates and traditional epistemologists, this amounts to a transformation of the subject, from justificatory concerns to descriptive-explanatory analysis.

By analogy with naturalized epistemology, there is also naturalized morality and naturalized aesthetics, both activities stemming from Darwin himself, and particularly from *The Descent of Man*, but recently enjoying a new lease of life under the general heading of sociobiology. As with naturalized epistemology, naturalized ethics and naturalized aesthetics take the scientific picture for granted; in the case of ethics and aesthetics this means mainly natural selection and modern genetic theory. It is then demonstrated how our ethical and aesthetic dispositions are founded in plausibly Darwinian strategies for survival and reproduction. As we shall see, where ethics and aes- thetics go beyond what these strategies dictate some, including Darwin himself, will throw up their hands in despair, and suggest a drawing-in of at least ethical ambition so as to come closer to our bio- logical nature and drives.

All this, both in its descriptive and, even more in its prescriptive guise, Socrates would clearly regard as an abrogation of philosophy's historic task, which is the determining of what is best and for the best. Philosophy, far from being limited to the descriptive and bounded by the naturalistic, is essentially (to repeat) a moral activity. It is worth at this point mentioning that Socrates' vision does have at least two mod- ern counterparts, Kant and Wittgenstein. While virtually no philoso- pher before this century would have remained content with the descriptive naturalism currently so fashionable, few before Kant would have seen philosophy as leading us to draw a clear distinction between science and morality, and to see the drawing of that distinc- tion as something morally motivated.

Yet this, surely, is the meaning of a critique of pure reason. It is to show the difference between natural science and other forms of discourse, including notably moral discourse. Drawing the line round scientific discourse is not, of course, a scientific activity, but something invoking philosophical considerations. *The Critique of Pure Reason* is thus definitely not continuous with science, nor indeed a reflection from within science. The derivation of causality, for example, or of substance are not intended as derivations from within Newtonian thought; they are supposed to state preconditions for any empirical thought whatever. And if critics—as some have done—suggest that in drawing them Kant was surreptitiously influenced by Newtonian considerations, rather than by attention to the essential preconditions of any empirical knowledge whatsoever, Kant himself would have regarded it as a disabling criticism of his enterprise.

When Kant comes on to deal with practical reason, he is at pains to show how the scientific model of causal explanation does not apply. In Socratic manner, he shows how practical discourse is subject to a different form of reasoning, a form in which the giving of intelligible reasons displaces the ultimately brute-factish quest for causal regularities. As we will later see with great clarity, Kant's approach to morality stands in stark contrast to the sociobiological demonstration of the source of moral behaviour and feeling as being the instinctive embedding in our genes and social practices of what conduces to our survival and reproduction. For Kant, by contrast, morality is divorced from instinct, psychology, or game-theoretic calculation, and is based in our nature as free, rational beings owing duties and respect to other free rational beings simply by virtue of their (and our) rationality. All this is quite independent of any ability on their part to harm or help us, and also quite independent of any empirically grounded feelings we or they have.

Although, as I will argue, Kant was wrong to divorce morality completely from its grounding in our biological and social nature, he was not wrong in thinking that it should not be constrained by it. The Kantian insight as to the theoretical separation of the empirical from the moral is correct, and the sociobiological attempt to deny or confuse this is as unavailing as is the evolutionary-epistemological attempt to defuse normative epistemological questions.

Like Kant, throughout his philosophical life Wittgenstein strove to articulate a sense of the limitations of science. In the introduction to the *Tractatus* he wrote, not it seems ironically, of how little had been done when all its problems had been solved. As the problems with which the *Tractatus* deals are those raised by the descriptions of natural science and by the demarcation of science from other activities, we can certainly understand the *Tractatus* in a Kantian spirit, an understanding which is to a degree confirmed in Engelmann's memoir of Wittgenstein. Much of Wittgenstein's later philosophy can legitimately be seen as a crusade against *scientism*, particularly in matters pertaining to the mind and human action. It is striking that in this crusade Wittgenstein—like Socrates—sticks with what he (and we) know best. He uses what some contemporary philosophers would dub 'folk psychology' as a basis from which to demonstrate the limits of scientific and philosophical psychology, but in considering the alleged weaknesses of folk psychology we should not forget that in practice we have no other way to describe or explain human action. Moving out of the folk psychological idioms (of will, intention, belief, understanding, pain, and so on) into something more scientific will mean that we cease to deal with human *action* (which is, after all, understood and conducted in the light of such notions), and shift our attention to neural and physiological events, for which action-guiding considerations and reasons do not and cannot arise.

As we will see, a similar though not so radical shift of subject-matter arises when we come to look at sociobiological accounts of selfishness and altruism (which is perhaps more worrying than the shift from folk psychology to neurophysiology because of the systematic ambiguities involved). However, Wittgenstein's own attitude to science, expressed near the end of his life, was a baleful one:

It isn't absurd to believe that the age of science and technology is the beginning of the end for humanity; that the idea of great progress is a delusion, along with the idea that the truth will ultimately be known; that there is nothing good or desirable about scientific knowledge and that mankind, in seeking it, is falling into a trap.[2]

Wittgenstein, no more than Kant or Socrates, would have regarded philosophy as in some sense secondary to science, or continuous with it. All three thinkers wish to establish a primary, fundamentally moral role for philosophy, and to use philosophy to curb the pretensions of science.

Having said this, though, we need once more to focus more precisely on Socrates' own complaint against Anaxagoras and his ilk. The complaint is that Anaxagoras makes no use of Mind when he explains the world (instead adducing causes like 'air and ether and water and many other absurdities'), and that is 'just about as inconsistent' as if he

---

[2] L. Wittgenstein, *Culture and Value* (Blackwell, Oxford, 1980), 56.

were first to say that the cause of everything Socrates does is Mind, and then go on to explain that he is in his prison in Athens because of the operations of his bones and sinews, making no mention of his conviction that he is acting in a way right and honourable. Socrates is thus taking himself as a microcosm of the world: the world, like the individual, is guided by Mind, and by what is for the best.

Far-fetched as Socrates as microcosm might seem today, those who would wish to explain human behaviour without reference to mind, and who would disparage explanations of action in terms of doing what is for the best as folk psychology, are making an analogous move in reverse. Rather than treating the macrocosm by analogy with the microcosm, they are treating the microcosm on analogy with the macrocosm. They are treating the microcosm (man) as if it were just part of the macrocosm, and guided and animated by the same principles. But this is surely misguided. Whatever we decide about our ultimate destination and origin, it remains the case that we, as human beings and as self-conscious agents, can question our standing in the world in a way no other part of nature can. This, indeed, is part of what 'acting for the best' comprises: raising questions about our relationship to the rest of nature and to each other. The normativity, the search for truth for its own sake, which this involves, engages us in types of considerations which are not found in the scientific descriptions and explanations, whether those of physics or of biology. So, anyway, I will argue in the course of this book.

The main thrust of my argument, which I will begin to develop in the next chapter, will be to suggest how self-conscious agency gives us goals and projects puzzling or even inexplicable on biological terms, as adumbrated in neo-Darwinianism. We can, however, immediately make a number of points of a more general nature, which will certainly be part of the context of our discussion. First, in practice science is and remains less certain than our pre-theoretical descriptions and explanations. Although we have already parted company with Quine on the nature of philosophy, the primacy of the pre-theoretical can be asserted on Quinean grounds: that revisionary schemes must always of their nature be partial. Thus, we can certainly conceive of science revising our view of, say, the size of the Sun or of its distance from the Earth. This has happened between Homer's time and our own, but this has not stopped much of the *Iliad* being intelligible to us, nor a clear sense in which we and Agamemnon may be said to inhabit the same world and be subject to similar vicissitudes of fate and mortality. Differences between us and the Argives there certainly are, but reports of these differences have been much exaggerated, and do not have that much to do with scientific revolutions. Nor, to take another sort of example, does knowledge of atomic structure and quantum theory do anything to impugn my sense that the table on which I write is flat, brown, and solid. If anything, it confirms this sense, because it goes some way to explaining how I come to have it. And it is as well that this should be so, because part of what we will be looking for from science is an explanation of the sense, not an undermining of it. It would be very difficult to see how human beings could continue their lives as human beings while not treating tables as flat, solid, and coloured. Thus the fact that at some level scientific theory says something other is not on a par with revised estimates of the Sun's size or distance. It remains a major philosophical task to reconcile the apparently divergent accounts of science and common sense here, which is in itself a testimony to the robustness of common sense in this area.

If it is hard to see science accomplishing a wholesale revision of commonsensical perceptions and beliefs about the empirical world, it is even harder to envisage science revising our attitude to practice, and the explanation we give to action in terms of agents' intentions and in terms of their being motivated to act for the best. This is because science itself is a practice, and because in choosing to do it at all and in doing it in particular ways, we will be subjecting ourselves to normative considerations. We will be having at the back of our mind the idea that, for various reasons, it is for the best that we engage in science, and that engaging in it in such a way is the best way to do it. In this sense, the decisions to do science and to do it in a particular way are on a par with Socrates' decision to stay in his prison cell, rather than let his bones and sinews (or genes) take him into exile. They are all decisions which cannot be given a scientific justification, and which demand a justification logically independent from anything we might discover in scientific accounts. It would then be, to say the least, self-defeating, if science—done in the best way and for the best motives, done in Socrates' terms because of Mind—were to tell us that Mind in this sense plays no part in human affairs, or that it is an illusion foisted on us by genetic working on quite other principles. This is a thought which will recur *en passant* throughout the book but while in the sixth chapter we will consider how far sociobiology undermines our sense that we can truly be acting for the best, in the fifth we will consider how far science itself may be held to undermine everyday acounts and perceptions of the physical world.

2

Immanent and Transcendent
Dimensions of Reason

In the last chapter I spoke of human beings as part of nature, but as, in a certain sense, standing apart from it. Following Socrates in Plato's *Phaedo*, I referred to the aspect of our nature by which we stand apart from the rest of the world as Mind, and Mind was analysed in terms of acting for the best. We need now to explore in more detail the relationship between our acting and thinking for the best, and the actual beliefs and practices we have, and in particular those with which we are born and which form the actual context within which we think and judge. What is the relationship? Is there not a danger that what we call our reasoning will simply be a copy of the standards and beliefs with which we are surrounded—and so not really independent of them, but only purportedly and mendaciously so? Or, if we avoid that danger, that we shall conclude that our beliefs and practices actually have no rational foundations or justifications, and so are left dangling in a sceptical limbo.

In a way, my theme in this chapter is a variant of Pascal's insistence on the multiple simultaneous realizations rational reflection brings about in us. We realize that our most basic beliefs and practices are assailable by sceptical doubts, but that these doubts are not ones anyone could live by. In living our lives we have to take so many things for granted, but in reflecting on what we take for granted we realize that it lacks the type of justification our reason seeks for it. We conclude, then, that there is something unreasonable in pushing demands for justification too far, in pushing them as far as the man who recognizes the force of sceptical arguments would have us go; but, following the Socratic impulse, there remains a niggling feeling that there is something terribly parochial in seeking to constrain rational inquiry within the framework set by that mixture of instinct and tradition which constitutes the framework within which we live. The feeling that there might be something irrational in curbing the pretensions of reason to criticize and conceivably even to undermine the basic premisses of our lives and practices grows into something more weighty when one reflects that some apparently basic premisses of our forefathers and of other cultures now stand revealed to us as false or even irrational.

Pascal says that no perfectly genuine sceptic ever existed. While this is true, and while this thought is the necessary counterpoint to the pretensions of reason to undermine a way of life completely, the difficulty is always to know whether in a specific case reason is overstepping its limits. At the same time, for many, the fact that no one has provided a generally acceptable solution to Descartes' dreaming argument, or to Humean scepticism about induction, or to modern fantasies about brains-in-vats is likely only to evoke Pascalian thoughts about the theoretical inutility and practical irrelevance of reason when it oversteps its limits.

These thoughts about reason are Pascalian not only in their recognition of the ambiguous nature of rational reflection and of its real or supposed limits. They also echo Pascal in seeing what might be thought of as the simultaneous *grandeur* and *faiblesse* of reason as stemming from the nature of rational reflection itself. Pascal believed that man's greatness consisted in the fact that he, unlike a tree, knows that he is wretched. Man is great because he is able to reflect on and become aware of his limitations. Pascal thought that this greatness through awareness of limitation had religious implications, but here I focus simply on the dialectic involved in our reasoning powers. On the one hand we have the ability and even the need to subject any belief or practice of which we become aware to rational scrutiny. On the other hand, in conducting rational scrutiny of beliefs or practices, we realize that our reasoning powers fairly quickly run out, and we are left with fundamental beliefs whose source appears to be the non-rational operation of instinct or tradition rather than reason itself and whose status as beliefs remain accordingly questionable. But, as Hume famously observed, on those things in which we cannot help believing in order to live our life—the external world, induction, the self, certain moral principles—nature becomes too strong for principle; we get on with our lives and in practice believe them anyway, whatever sceptical scruples remain at the theoretical level.

As a matter of fact, even if we have reflective and sceptical tendencies by virtue of our nature as self-conscious beings, and even if we agree with Socrates that in a certain sense the examined life is the one most worth living, we are still creatures of nature. As such we have

instincts and biological drives which impose constraints on our conduct and, indirectly, on the way we perceive and think about the
world. While the would-be moral reformer might on rational grounds
and in the spirit of a Plato or a Bentham propose an as-yet untried
form of existence as the best for man, human beings are not infinitely
malleable, nor can happiness be brought about by the untrammelled
exercise of individual choice (however rational in intention) if the
choices involved do too much violence to our nature. Analogously,
Hume's sceptical doubts vanished when he turned from his study to
the backgammon table, and one wonders if Newton himself failed to
feel a tension between what his *Opticks* taught and his senses told him,
and where, if such a tension were felt, the victory lay as far as everyday existence went. Or, to put this point another way, is there any
physicist of today who teaches his children the language of physics
before they have mastered the vocabulary and world-view of instinct
and common sense?

When we look at our behaviour, we can see that once again instinct
seems at critical moments to present us with essential modes of
response. These modes present themselves as impulses or imperatives
which any amount of thought and reflection can do little to weaken. A
mother in defending her children against attack or a man in responding to insult or invasion of his territory can surely be seen as repeating
primeval patterns of behaviour deeply rooted not just in our species,
but in ways of life far more primitive than our own. And in one way it
is well that we draw on this pool of unthinking, natural response. It is
another symptom of the fact that we are natural beings, with a nature
partially formed by nature underlying our choices and decisions. It is
a sign of our lack of absolute malleability and, at the same time of the
inheritance which defines individuals as individual human beings, and
not as unformed plaster awaiting the hands of some would-be sculp-
tor of human nature.

'The mother who fights in fury to defend her child is—one would
hope!—obeying the call of instinct, not the advice of some maternal
guidance leaflet.' Bruce Chatwin's eloquent paean in praise of human
instinctive behaviour touches on a crucial point.[1] The call of instinct of
the women fighting for her child, which is manifested equally (as
Chatwin avers) in the fighting behaviour of young men, presents itself
to most people as an absolute, an imperative unconditioned by reason,
not done for the sake of some other end. The mother defending her

---

[1] Bruce Chatwin, *The Songlines* (Picador, London, 1988), 240.

child does not reason that this is what she ought to do; at least she does not if her feelings are healthy and not sicklied o'er with the pale cast of thought. And, as will emerge from a quick survey of modern applied moral philosophy, infanticide, lying, 'deep' ecology, absolute pacifism, forcible population reduction—and their opposites—can all be defended on rational (or 'rational') grounds. It all depends on your initial starting point. In the case, though, of the mother and her child, there is no starting point of principle, no reasoning from principles, there is just the response.

The contrast we are drawing between the imperatival, categorical demands of instinct and the apparently tentative, less than basic conclusions of reasoning processes raises a question as to the nature of reason. If reasoning is what reason goes in for, it can seem less a source of ultimate ends (which might be supplied by our biology, our instincts, or even our station in life), than a means by which we can fulfil our ultimate ends or bring them into harmonious conjunction in cases where we face conflicts of ends. The contrast between instinct and inherited or acquired duties, on the one hand, and reasoning processes on the other has notoriously led many to follow Hume in thinking of reason in purely instrumental terms: reason can deliberate about means, and refine or improve them, but about ends it must be silent. And where basic beliefs are concerned, too, nature is too strong for principle and breaks the force of all sceptical arguments.[2] Once again, we get our basic beliefs from a non-rational source.

It would indeed be hard to quarrel with the view that a great deal of explicit reasoning activity is reasoning about means and what Aristotle would call intermediate ends—and this both in matters of practice and of belief. In physics and cosmology, for example, we do not spend much time examining the beliefs that there is an external world apart from us, that it has existed for a long time and that it manifests and will continue to manifest causal and other regularities. What we do could be seen as the attempt to work out the implications of these ideas given the phenomena we experience. Analogously, in ethics and in politics disagreement over fundamentals often becomes relatively unimportant when participants in a discussion begin attempting to devise practical proposals to deal with some admitted problem of mutual concern. And thus the suspicion grows that the reason why reasoning about means is far more common than reasoning about ends is not just because the latter is far more difficult. Might it not be that the latter is

---

[2] Cf. D. Hume, *Treatise of Human Nature*, ed. L. Selby-Bigge (Clarendon Press, Oxford, 1888), 187.

far more difficult just because in the end, ends—first principles—are not given by reason at all, but by nature, instinct, tradition, form of life, or some other basically non-rational source? (Or a source, which if rational is rational in virtue of what Hegel thought of as reason's cunning or what followers of Adam Smith might characterize as an invisible hand, rather than by anything an individual rational inquirer could necessarily perceive as reasonable at any given time?)

The idea that reasoning proper is only ever about means and never about ends is, of course, too crude as countless critics of Hume and his followers in this thesis have pointed out. For one thing, ends adopted by one individual or group may in practice conflict and the individual or group be forced to reflect on the relative desirability of the various ends they are proposing to themselves. For another, the very fact that a given end requires such-and-such means can lead people to question the desirability of the end. (Many of those who have initially been attracted by Platonic ideals of strict equality of opportunity in education have come to question the validity of those ideals once they have realized that only the Platonic means of taking children away from their parents at birth would have any hope of achieving them.) Then again, reasoning in the broad sense of having the requisite intelligence and sensitivity is needed in order to perceive that such-and-such an action is what courage or some other virtue requires in given circumstances. It is not always immediately obvious what our end should be; reasoning, then, is required in the determination of ends just as much as in the choosing of the best means to reach our chosen ends.

The considerations of the last paragraph strongly suggest that in practical matters ends and means are far more closely intertwined than is realized by those who think of reasoning purely in terms of means. The consequence is that it is impossible to restrict the operation of reason to the discovery of means alone; reasoning is also involved in the determination and choice of ends. Analogous considerations apply in the epistemological case. Even if, as I argued earlier, we do not spend time in physics or cosmology arguing about the existence of the external world, or its longevity, physics and cosmology have had a huge impact on our conceptions of the nature of that world and of its actual time span. In this way, a study initially undertaken within certain presuppositions has changed our ideas about those presuppositions in ways which could not have been foreseen when human beings first embarked on the study.

However, even though we would be right to reject a crude demarcation between the rational and the non-rational in terms of proximate and ultimate ends or assumptions, it remains true that our ideas about what it is reasonable to hold in matters of belief and of practice do not float free of our systems of belief and practice. They are embedded within them, required by them, and are not justifiable outside them. Even giving reason a wider scope than reasoning about means, it remains true that Cartesian scepticism about dreaming, or Humean doubts about induction or about the rationality of ultimate practical ends cannot be settled by us without assuming the validity of the very practices such scepticism is intended to cast doubt on. And the fact is that as human beings we have no option but to accept the validity of our practices, and to take their deliverances and standards as the criterion for the reasonable. In terms which would have made sense to Cardinal Newman, reason in the sense of that which determines what is to count as reasonable should be seen as abstracted from the flow of life, and not as standing outside our life.

Reasons of an explicit sort, then, should as Michael Oakeshott argues, be seen as an abridgement of our practices, which are themselves grounded in our instincts and traditions.[3] When we ask—as Hume does—whether it is more reasonable to prefer the scratching of our finger to the destruction of the whole world, or when we seek, like Descartes, to establish whether or not we are dreaming, we seem to be asking for principles on which to found our beliefs and practices. Examination of the questions and of the principles which might emerge, though, suggests that these and similar sceptical questions can be answered only given we accept the frameworks of instinct and tradition which we inherit. Given human life, desire, and community, it is naturally preferable not to destroy the whole world, but as a number of ethical thinkers from our tradition (including Schopenhauer and Tolstoy) have suggested, there need from an absolute point of view be nothing sacrosanct or abstractly compelling about seeking the preservation and continuation of the human race. (This view would appear ultimately to form the basis of the Buddhist four noble truths.) In everyday life, it is only in very abnormal conditions that there is any doubt as to whether one is awake or dreaming. Not even abnormal circumstances can make one's really being a brain in a vat seem plausible. Nevertheless, as the history of philosophy teaches, scepticism and solipsism remain constant temptations to the rational mind, and rational argument alone cannot dispose of them, unless one is prepared to start from an initial acceptance of the reasonableness of animal belief and human practice.

[3] Cf. M. Oakeshott, *Rationalism in Politics* (Methuen, London, 1967), 120.

All our reasoning, then, starts from the fact that we are are embodied beings—rational animals, but animals none the less—with a life constrained by a world we did not create. In so far as philosophy tends to undermine confidence in this *fundamentum inconcussum* of our thought and existence, it is superfluous and deserves the dismissals it regularly attracts. As I will argue in the next chapter, the very fact of self-consciousness—*pace* Descartes—suggests that we are members of a human community in a world existing apart from us. But saying this will dispose only of the most extreme forms of hyperbolic doubt. It will do little to underwrite any *specific* beliefs we have about that world, or about what is truly valuable in the community. We should, then, beware of pushing any simple repudiation of sceptical questioning too far to the extent that we deprive ourselves of any scope in the area of rational justification and inquiry, just as, in Chapter 5, I shall argue that scientific speculation should not be allowed to undermine initial confidence in the validity of our 'folk-psychological' perspective on the world.

In the view of Pascal, reason's last step is the recognition that there are an infinite number of things beyond it.[4] Pascal may have been thinking there of truths of a transcendent sort; what I am suggesting is that there is equally a sense in which the most basic and mundane truths are beyond the power of reason to establish, or indeed, to undermine. There can actually be something unreasonable, something only half-serious, about engaging in the sort of fundamental sceptical doubt about our basic beliefs and practices which the practice of the sceptic continually belies. To admit the limits of reason in both the criticism and the justification of our beliefs and practices is not unreasonable. What is unreasonable is to think that reason can be effectively employed outside the context of belief and practice in which reasons and reasoning have their home.

An Argument from Utility

What has so far been said about reason and reasoning has no transcendent implications. Human beings are being seen as creatures of nature, with certain naturally and culturally given dispositions and beliefs. Deeply embedded beliefs and practices—about the external world, about induction, about the imperative need an individual has to sur-

---

[4] B. Pascal, *Pensées*, trans. A. J. Krailsheimer (Penguin, Harmondsworth, 1966), no. 188.

vive and to protect his or her young—can plausibly be seen as having been embedded because they have promoted the survival of individuals who held them, and indirectly of the species as a whole. We can also speculate that the unavoidability of what Hume refers to as animal belief in body and probability and the rest, and the unthinking and immediate nature of many of our responses to what we perceive as threatening to ourselves or our dependants is due to nature's having done the thinking for us over millennia. These are the beliefs and responses which those who have survived best in the past have possessed. These are the beliefs and responses which have been demanded by and have themselves fostered the flow of human life, expressed and reinforced through a dialectic of instinct and tradition. As such, they make absolute demands on us, demands which reasoning is, in the main, powerless to weaken, however much it might make sceptical noises. Nature is, as Hume says, too strong for principle.

In the light of what has been said, reason has a wider role than merely discovering means to non-rationally assessable ends or just working out the detailed implications of a largely inherited picture of the world. By reasoning we can both develop our systems of ends and our picture of the world. Nevertheless reason and reasoning may still be seen as playing a largely instrumental role within the flow of life. At any given time there is much which simply has to be accepted because it is demanded by life, beliefs, and dispositions which are neither justifiable by reason in the abstract nor overthrowable by reason. Reasonable values and beliefs are always those which in some senses conduce to life, assist us in our commerce with nature and with our fellows.

I am well aware of the vagueness and ambiguity of the idea of values and beliefs being conducive to life. Very likely what Nietzsche might see as satisfying such a description would be quite different from what some representative of the deep ecology movement might see in such terms. Some readers, too, will be alarmed by the mere mention of Nietzsche in this context; what is from one man's point of view conducive to (his) life might look to another like the suppression of his rights and the undermining of his cherished projects. Speaking of life as the foundation of value inevitably raises the spectre of the will-to-power and of social Darwinism. Nevertheless there remains some plausibility in the idea that at a certain level some beliefs and dispositions are simply forced on us by the facts of life. If this is so, then it need not be the case that the rationale for the beliefs and dispositions in question be utterly transparent to those on whom they are, so to speak, forced. They may find themselves, like Hume, just accepting the cogency of belief in body or inductive inference without being able to justify these beliefs to themselves. It may also be the case, as Hayek has argued, that we are not always clear just what our useful dispositions are, or why they are useful. If they have been selected because they conduce to the well-being of those who hold them, what matters is whether those who hold them act in the appropriate way, not whether they are clear just what the dispositions are or why they are useful.

So the possibility has to be admitted that, in addition to the Humean animal beliefs in basic metaphysical realities, myths and religions could be the means by which life-conducive dispositions are passed on from one generation to another, even though individual rational minds may fail to understand the function or validity of the beliefs in question. That speculation of this sort is the stock-in-trade of sociobiology is not enough to discredit it. *If*, as many post-Humean philosophers are prepared to admit, epistemological scepticism and solipsism is rationally undefeatable, and defeated in practice only by pressing biological and social needs, why should not other rationally indefensible myths at times play an equally salutary role in the lives and cultures of those societies which foster them?

Without denying that what is to count as reasonable in matters of belief and practice is often to be seen in terms of the embedding of a belief or practice in a way of life, against the line of thought developed in the last four paragraphs, I want now to begin to argue that our reasoning ability itself cannot be constrained by the useful. Nor is reason simply to acquiesce meekly in what is given and abandon all thought of justification or improvement of even our most basic beliefs and principles. There are powerful reasons for arguing against such intellectual passivity, reasons which arise from the nature of human beings as self-conscious agents, even if we wish to maintain that self-consciousness and what I will call reasoning were initially selected for the evolutionary advantage they brought to our species. What is crucially at issue here is not how human self-consciousness might have come about, but what its significance for its possessor is once it has come about.

What do we mean by reason and reasoning? There are, to be sure, many aspects of human life which have at one time or another been seen in terms of reason. One main distinction which has to be drawn immediately is that between reason as content or product and reason as activity or process. When, as we have done ourselves, we talk of beliefs or practices which are reasonable, we first mean beliefs and practices which have emerged as reasonable, or which have survived the scrutiny of reasoning. An extended sense of the reasonable is that described by Hegel as the cunning of reason and by Hayek, following Adam Smith, as the work of the invisible hand: where practices and beliefs survive in a community because they are the beliefs or practices of those who are successful (and so eventually of a successful community) despite the fact that those who hold or practise them may do so subjectively for quite other reasons (or may, in the extreme case, not even be aware that they have the belief or disposition in question).

Now while reason as content is a crucial aspect of reason, and one which cannot be underemphasized, even when we consider the activity of reasoning (for how else can reason work without a framework delimiting what the reasonable is?), it is the activity of reasoning which I now want to focus on. In the *Politics* (1253a7–17), Aristotle speaks of the real difference between man and animals being the perception on the part of humans of good and evil and of the just and the unjust. He connects this to the sharing of a *common view* on such matters (which is what constitutes a household or a state), and on the possession by humans of speech (logos) as opposed to mere voice or expression. In terms of the distinction I am drawing, Aristotle's common view will be the reasonable, the content of reason. Speech (logos) is what the activity of reason is. (While we might be reminded here of the Platonic account of thought as the soul's dialogue with itself, I prefer to concentrate on the Aristotelian account, because, as will emerge, particularly in the next chapter, his invocation of a common view as founding the reasonable is closer to the truth than the Platonic account of reason's content.)

Both the idea of speech and that of the dialogue of the soul with itself suggest a reflective questioning of what is initially asserted or believed. It is this reflective questioning which is the essence of reasoning activity, the looking for reasons for what is asserted or believed, the testing of what is asserted or believed in the light of what is really true or good. It is, I will now suggest, an essential concomitant of self-consciousness; it is, indeed, part of what self-consciousness amounts to.

In being conscious of myself as myself, I see myself as separate from what is not myself. In being conscious, a being reacts to the world with feeling, with pleasure and pain, and responds on the basis of felt needs. Self-consciousness, though, is something over and above the sensitivity and feeling implied by consciousness. As self-conscious I do not simply have pleasures, pains, experiences, and needs and react to them: I am also aware that I have them, that there is an 'I' which is a subject of these experiences and which is a possessor of needs, experiences, beliefs, and dispositions. In being or becoming aware of experiences, dispositions, and beliefs as mine, I am at the same time aware that there is or might be something which is not me or mine, something which my experiences and beliefs are of, something towards which my dispositions are directed. And this essential contrast of my mental universe with a world which is not me and not mine immediately raises the question for me as to the adequacy of my mental universe as a representation of the world which is not me or mine.

A self-conscious person, then, does not simply have beliefs or dispositions, does not simply engage in practices of various sorts, does not just respond to or suffer the world. He or she is aware that he or she has beliefs, practices, dispositions, and the rest. It is this awareness of myself as a subject of experience, as a holder of beliefs, and an engager in practices which constitutes my self-consciousness. A conscious animal might be a knower, and we might extend the epithet 'knower' to machines if they receive information from the world and modify their responses accordingly. But only a self-conscious being knows that he is a knower. This knowledge, I want to suggest, introduces a dimension to reasoning which we have not yet examined.

In becoming aware that I am a knower or believer, and analogously that I am an agent, I cannot but begin to understand that my beliefs and dispositions are just that—mine; and that being mine, they are not necessarily in tune with those of others or with the world. In other words, they are not necessarily true or correct; they are but one route through a world apart from me. Once the soul becomes aware of the nature and condition of its own knowledge and dispositions it is led to postulate an absolute standard against which its mental map is to be judged. And so we get an aspect of reason and reasoning quite different from our earlier conception of reason as the instrument and slave of life. Its nature is also to be the critic of life and the passions and that which we cannot but believe. Many philosophers take the real lesson of scepticism to be the revelation of the simultaneous groundlessness and necessity of our knowing and our practices. But equally significant is the way scepticism is a striking symptom of what we might call the vertical aspect of reason. At any moment the demands of life's flow can be held up by us to scrutiny, we can step outside the steady stream of judgements and practical decisions we are continually making so as to see if they satisfy some standard not limited by the limits of our life or cognitive powers. We can, as it were, step outside our cognitive and practical frameworks and question the validity of the frameworks themselves, or at least of fundamental aspects of them.

The lesson of philosophical scepticism is that such transcendence of the everyday and the commonplace is not confined to cases such as that in which Einstein questioned the absoluteness of simultaneity. Einstein was certainly questioning something fundamental to the conceptual scheme of his contemporaries, but not sceptically. He was proposing to replace one plank of the ship of physics with another. The sceptic, by contrast, has nothing to put in the place of what he attacks. But he can raise questions about it none the less.

Because he has nothing to replace what he attacks, and because he cannot live without it, the sceptic's questioning is, as we have seen, unavailing and in a sense unreasonable. He takes it upon himself to judge life and its demands, to point to the groundlessness of our beliefs and the arbitrariness and contingency of our practices. From the point of view of his questioning, the sceptic demonstrates the infirmity of reason, but in so doing, he also demonstrates its ability to step outside the particular content of any given set of beliefs or practices, and to assume the role of transcendent judge. It is, of course, our nature as self-conscious, reflective agents which enables us to assume this role. But when we reach the limits of our beliefs or practices it is not a role we can do anything with, or at least not anything which will itself survive the scrutiny of transcendent reason. The reason for this is that whatever we might put in place of what our questioning has undermined will be vulnerable in its turn to further rational criticism.

Some would draw from this tale the moral that reason should stay firmly within the limits of whatever is required by the flow of life, that reason should recognize its enslavement to the passions and its subservience to animal nature. One difficulty with this approach is that it is not obvious just what these requirements might be. Reason itself will have to be called on to determine them. Moreover reason itself is a product of nature and it is hard without special pleading to say why following reason wherever it might lead is not staying within the confines of animal nature. In any case, any injunctions to curb rational questioning are bound to be unavailing. Such questioning is bound to arise once self-consciousness has made its appearance in our species, being an inevitable concomitant of the self-conscious mind.

What I suggest is that we need to admit both the infirmity of reason and its transcendence, its infirmity or limitation regarding content and its transcendence as process, its need to find truths and standards which stand absolutely justified and its inability to discover any. Recognizing this predicament should lead us to avoid two opposite, but related, pitfalls. The one is that of the rationalist, who recognizes the limitations of our cognitive and ethical inheritances and the infinitely questioning and critical aspect of reason, but fails to appreciate the value of these inheritances and (what comes to the same thing) our inability to do without them. The other pitfall is represented by the philosophical conservative who recognizes the limits of reason from a constructive point of view, but whose reaction to these limitations is to seek to curb reason's questioning, critical aspect. In the case of reason, neither conservatism nor constructivism is going to work. The one unavailing seeks arbitrarily to limit the scope of rational inquiry, arbitrarily because the conservative will make full use of reason as an instrument, without realizing that the instrument is naturally geared to reflect on anything that comes in its path. The other, though, while understanding the limitless ambitions of reason does not understand that we, as human beings, are always *in mediis rebus*, and can neither escape life's demands (through philosophical scepticism), nor rewrite them by making some form of new and rationally underwritten fresh start in either belief or practice.

The simultaneous recognition of both the demands and the limits of human reason receives concrete expression, and some would say, its natural fulfilment in religion. From the perspective of pure reason morality might seem to be based on merely human institutions and sentiments, but through the experience of guilt its demands nevertheless present themselves as unconditionally binding on us. Religion solves or attempts to solve this dilemma by assuring the believer that the demands of morality are indeed absolute and transcendent, reflecting the will of God rather than simply serving human utility. Similarly our beliefs about the natural world and our science may appear constrained by the limits of our minds and perceptual faculties. But in religion we gain an assurance that our knowledge does stem from God, who is the source of all truth, and that we are not completely limited by the finiteness of our minds.

What I am suggesting is that there is in our nature as self-conscious but finite beings an ontological tension which naturally expresses itself religiously: that religion in its intimations of human limits and their overcoming mirrors very precisely our nature as thinkers and reflective agents who are nevertheless constrained in practice by the finitude of our powers. From the religious standpoint, humanistic approaches to belief and practice will be inevitably flawed, reflecting either an intelligence-denying conservatism or a hubristic rationalism, whose outcome is only too likely to be ethically and practically disastrous.

Religious belief does not simply understand and express our inter-connected limitation and transcendence. It also sees our drive to some-thing transcending human powers as reflected in the fabric of the universe, our yearning as for something which actually exists. Religious belief implies that there is a transcendent aspect to reality and that we are part of and related to this aspect. The non-religious will, of course, reject this, even where they find themselves in sympa-thy with the religious *via media* between inflexible conservatism and hubristic rationalism. The religious will no doubt point to what is widely regarded—even by materialists—to be a mystery, namely the existence of consciousness and self-consciousness. Certainly science has difficulty in accounting for the appearance and nature of conscious and self-conscious processes, in accounting for the way in which bio-chemical processes lead in animals and humans to a felt awareness of the world and the body and, in humans at least, to a reflexive aware-ness of oneself as an agent and a thinker. It will be natural for the religious to interpret this emergence of consciousness and self-consciousness as revelatory of something deep in the universe, some-thing inexplicable by physics, something behind the material face of the world.

The religious will be encouraged in this line of thought by the realization, which we have stressed, that it is through the fact of self-consciousness that we become aware of our questioning transcen-dence of the given, and also of our inability to get completely beyond what we have been given. In other words, that by which we become aware of our dialectical relationship with our cognitive and ethical inheritance is that which is also a sign of the source of resolution of the conflict in a reality behind or beyond the material world. In being self-conscious we inevitably assess our beliefs and practices for their truth and validity. We find in our experience no absolute criteria by which we can satisfy this drive for assessment. But self-consciousness itself intimates a notion of criteria more absolute than the perpetual contin-gency of the material, physical world, criteria deriving from a world where, according to the religious, absolute truth and absolute good-ness exist.

My view of the transcendence of self-consciousness is one I share with certain critical theorists, who also see consciousness (or, in the terms of this essay, self-consciousness) projecting us beyond the pres-ent. It does so by forcing on us the realization that our path through the world is but one possible path, and that our beliefs and practices represent only one route through the world, a route which is inevitably clouded by the partiality of our perspective and the physical and mental limitations of our faculties. As one critical theorist puts it: 'since it possesses a utopian truth-content which projects beyond the limits of the present, consciousness is transcendent.'[5]

There is, though, a crucial difference between the position espoused here and that of the critical theorists. There is no reason to believe in utopia as a solution to the dilemma posed by self-consciousness. Indeed it could not be: any society on this earth would be constrained by its own perspective and institutions, perspectives and institutions which would themselves come under rational scrutiny and criticism from the reflective intelligences of their own time. This objection to what Voegelin speaks of as the immanentizing of the transcendent is so far a purely theoretical one. History, including recent history, has painted a far bleaker picture of actual attempts to produce utopias. As I shall argue in later chapters, conservatism is nearer to the truth about established values and institutions than critical theory: conservatism sees them not as the cloak behind which power structures hide (or not simply as that), but as the product of decades of quasi-natural selection, whose overthrow is likely to incur considerable cost in unforeseeable directions. Moreover, the conservative is quite correct to emphasize the public character of self-conscious inquiry, and its basis in established values and institutions, something to which we will turn in the next chapter.

Religion, then, is right to postulate that the goal of the transcendent aspect of our reasoning powers will not be found on this earth, now or in the future. Anything which we came across on the earth would lack the absoluteness and unconditionality necessary to silence scepticism and criticism. Even if we showed that life could not go on without certain beliefs, that certain practices and only those practices fulfilled some deep sense of human need, these demonstrations would only lead to a provisional absoluteness. They would show us only that if we want to live and fulfil our natures certain things are forced on us. But this would not by itself silence doubts about, say, the justification of induction or about the absolute value of human life. As is shown by the examples of Hume and Popper, on the one hand, and by Schopenhauer and Tolstoy, on the other, it is perfectly possible to entertain doubts on both these issues without evident irrationality or

⁵ S. Beahabib, *Critique, Norm and Utopia* (Columbia University Press, 1986), 4.

insanity, however hard it may be to live out the doubts in practice. And the same no doubt goes for scepticism about other beliefs and practices which are constitutive of human life.

It may, then, well be that, as Hegel says, man can respect himself—and accord to individual human life that sacredness and dignity secular theorists find so hard to establish—only if he is aware of a higher being, to whom human life is referred and from whom it is seen as coming. It may also be, as Descartes argued, that our claim to have found truth in any area ultimately requires the assumption that our reason is a mirror of the mind of God. (Secularists tend to get round this problem by claiming—implausibly and inconsistently, the religious might feel—to be fallibilists about all our beliefs.) While reflections of this sort might lead us to a sense of the fragility of the ground on which we stand cognitively and ethically, it does not, though, make religion any more believable. In particular, as the arguments of Kant above all have shown, we cannot get to God by means of any form of argumentation which seeks to prove the existence of God by means of discursive argument from empirical or quasi-empirical premisses. In much the same way that a political utopia is no solution to the dilemma of our consciousness—being itself yet another empirical and hence contingent phenomenon—the end point of any argument from cause or design is itself going to be an entity of which further causal questions can be asked. The only way out of this regress is for the believer to stipulate that God is a logically necessary being, a being whose essence it is to exist, and who therefore cannot but exist, but as numerous discussions of the ontological argument have shown, it is doubtful that the concept of a logically necessary being is coherent.

Nevertheless, even if natural theology is intellectually unsatisfactory, and the credal structures of dogmatic religions even more so, I want to insist that in our self-conscious search for the true and the good absolutely speaking, we do have intimations of a realm of absolute value and truth, a realm to which religions do point, however unsatisfactorily. One could indeed suggest that any religious system, in so far as it is determinate and expressed in human language, is bound to leave the transcendent consciousness with as many problems as answers: why God or Brahma is as he or it is, why he disports himself in this way rather than that, and so on. As Karl Barth and other modern theologians have frequently complained, religious creeds and systems too often make of God a being among beings rather than the pure ground of existence, that totally other, beyond all religious and beyond all human determinations and formulations, which alone could put a stop to all questioning.

What I have been calling our transcendent reason or self-consciousness manifests itself most directly when we seek truth as that which has no interest for us. Here we act out the dynamic suggested by our conception of ourselves as believers or thinkers. We realize that our perspective on the world and our set of beliefs form only one view of the world, and one not necessarily correct. Thereby we begin the critical process which leads us to attempt to revise and improve our beliefs, for we cannot at the one time believe something and doubt its truth. The very fact of being self-conscious about our beliefs, of being in a full sense believers, then initiates a process in which we search for what is true because it is true, rather than because it serves some interest of ours.

Later on I shall develop the implications for us of this striving for truth for its own sake, and I shall then pursue an analogous thought with regard to goodness. But I will close this chapter by saying a little about the bearing aesthetic sense has on our nature as self-conscious beings.

In aesthetic experience we encounter spatio-temporal objects, music, painting, landscape, and the rest. But (as Kant argued) we encounter them without interest and for their own sake; in a sense, we approach them as abstracted from the flow of life, in and for themselves. Looked at in this light, G. K. Chesterton's claim that every true artist feels 'consciously or unconsciously that he is touching transcendental truths, that his images are shadows of things seen through the veil' can begin to make sense.[6] But it is in experiencing worldly things *sub specie aeternitatis*, in and for themselves, that we gain the sense of an eternal world behind the aesthetic object. Unlike much religion, aesthetic appreciation does not seek to deny or play down our embodiment or our temporal existence. Nor is it encumbered with unbelievable dogmas and thought-forms. As I will argue in my penultimate chapter, it is in aesthetic experience that our self-consciousness can best find some quietus in its restless movement between enmeshment in the flow of life and the search for something transcendent to justify its beliefs and practices absolutely. Whether, though, this provides any justification of religious or Platonic sentiment must remain a moot point, given that in aesthetic experience both perceiver and object remain firmly embodied and enmeshed in the natural material world.

---

[6] G. K. Chesterton, *The Everlasting Man* (Hodder and Stoughton, London, 1925), 120.

3

Self-Conscious Belief

The dialectic we began to examine in Chapter 2 was that between the reflectiveness of self-conscious belief and the embedding of our thought and practice in unconscious instincts and traditions. We may question our instinctual or traditional inheritance of thought and practice in any way: theoretically there are no limits to reflective questioning of our practices. This is one of the lessons we learn from the history of sceptical thought. At the same time, though, our existence as agents, embodied in a particular way and living at a particular time, puts limits on the extent of what might be called effective questioning. Reflective reason outstrips the constraints of unthinking practice, only for the reasoner to be pulled back to earth by his need to engage in practice.

I want now to say a little more about the nature of reflective reason and the epistemological and social conditions which might foster its development. In pursuing this question some interesting aspects of the nature of reason itself will emerge.

I wrote earlier (Chapter 2, pp. 21 ff.) of the self-conscious believer being aware of having beliefs and dispositions, of knowing that he is a knower, and that he is taking one possible route through a world which exists apart from him. A crucial contrast was drawn between this fully aware and self-conscious agent, on the one hand, and a merely conscious, but not self-conscious being. Both the contrast and the description might seem to suggest that the model or type of the self-conscious believer is Descartes withdrawing from the hubbub of the world and the Thirty Years War into his stove-heated room, or perhaps Hume enveloped in his studious mists away from the laughter and chatter of the backgammon tables. Indeed, these classic images of philosophical reflection are not entirely unilluminating. Philosophical reflection does have an inbuilt tendency to solitariness and even to solipsism. It is, crucially, the reflection of an individual, who for the moment stands apart from the world and the practices he is contemplating. It does involve a quest for self-knowledge of a sort, a quest for knowledge of what I know and what I am. But, as I shall now attempt to show, the solipsistic stance is not the starting-point of self-consciousness, but a stage some distance along the philosophical road. Self-conscious reflection might lead one to a solipsistic conclusion; it might lead too to the thought that one knows oneself and one's own mind better than anything else. But it would be quite misleading to think of introspective self-knowledge as one's actual epistemic starting-point, or to think that it can actually exist without personal existence in an objective world, among other believers and speakers. If this is right, then it will be mistaken to draw solipsistic conclusions from an analysis of one's self-consciousness or to attempt to found knowledge of the external world on introspectively available data. For my sense of myself as a believer or a reasoner, and as a possessor of introspectively examinable data will depend on a prior recognition of other people in a world which has an existence independent of me or of my perceptions of it.

The question which is being posed here is a question which was not put in the previous chapter. There we simply assumed that each of us was in the position of being a self-conscious agent, a self-conscious believer. Arguably it is taking this as an unquestioned assumption which has classically led philosophers into scepticism and solipsism, because once what is involved in coming to a recognition of oneself as self-conscious becomes apparent, it follows that solipsism and at least certain extreme types of scepticism must be false. What is involved in becoming self-conscious? How do I, or did I, become aware of myself as a believer?

In order to answer these questions, I will first distinguish between consciousness and self-consciousness, and it will then become clear that to be a believer in the full sense, self-consciousness is required. Consciousness involves reacting to stimuli and feeling stimuli. An unconscious being, like a photo-electric cell or a thermometer, may be programmed to react to certain stimuli. As a result of its reactions, certain other things may happen: doors may open, heaters may be activated, but there is no question of the cell or the thermometer actually experiencing a shadow or feeling some level of cold. Animals and even some more primitive living organisms are more than reactors to stimuli: they also *feel* stimuli, as pleasure and pain, for example.

It will become important in what follows to realize that consciousness in humans and animals is not simply a matter of information-processing. For information-processing, indeed, there may be no significant advantage to be gained through being conscious. If all we are interested in is the representation of impersonally specified objects and events, and the manipulation of such representations, it becomes hard to see what advantage there is in being conscious, over and above performing the relevant calculations. Computers can do as much, and often more speedily and efficiently than we can.

As Gerald Edelman has argued persuasively and at length, biological memory, whether in animals or human, is not to be regarded as a replica or trace coded to represent some object.[1] It is rather 'an open-ended set of connections between subjects and a rich texture of previous knowledge' which cannot be adequately represented by what he calls the 'impoverished' language of computer science.[2] Even more significant, the parts of the brain involved in consciousness are the limbic-brain stem and the thalamocortical system, which work together in conjunction with re-entrant circuitry to produce conscious experience. While the thalamocortical system is involved in conjoining motor behaviour to the categorization of objects or events, the limbic-brain stem system is above all a vague system, evolved to regulate various bodily functions, including digestion, sleep, and sex.

Edelman's hypothesis is that the two systems, the limbic-brain stem and the thalamocortical system work together in us and higher animals to produce what he calls scenes, that is, conscious episodes in which present events are imbued with the values accorded to similar events in the past history of the subjects. As he puts it, 'the salience of an event is determined not only by its position and energy in the physical world but also by the relative value it has been accorded in the past history of the individual animal as the result of learning'.[3] The evolutionary advantage of consciousness derives not from what might be called the 'cold' cognition of information-processing, which, as we just argued, need not be conscious at all. It derives from the tone or feel or value accorded in consciousness to a perceived scene. What Edelman calls primary consciousness imbues what is perceived with values determined by the animal's past history and experience. It is plausible to suppose that the importation of digests of past evaluation into an animal's present will enable it to generalize from past learning far more quickly and relevantly than an animal lacking this ability.

This primary consciousness lacks any notion of an experiencer or self. It is as if the animal is in a darkened room, able to shine a beam of light on what strikes it;[4] the beam does not simply illuminate what is

[1] G. Edelman, *Bright Air, Brilliant Fire* (Penguin, Harmondsworth, 1994).
[2] Ibid. 238.   [3] Ibid. 118.   [4] Cf. ibid. 120.

perceived, it endows it with meaning and value derived from the animal's past. But all else, including any explicit grasp of the self, or of the animal's past or future is unthematized. It may bear on the present, but in an unrealized way.

Thus, to take an example of Mary Warnock's,[5] a horse may have had a bad experience in the past when being loaded into a trailer. This bad experience conditions its present perception and attitude on being loaded now. This type of event will be familiar to anyone who has had dealings with horses or other animals, and they will also know that the way to overcome a resistance is to set up new, non-threatening associations in the animal's mind. But even though felt pleasure and pain, including the sense of past pleasures and pains deriving from similar past experiences, are crucial aspects of an animal's reception of a stimulus and its response to the stimulus, it does not in itself involve the animal having a conception of itself as a being who feels pleasure or pain or who has had a past or who will have a future. It is this conception of oneself as a perceiver, as a subject of pleasure or pain, and as an agent in the world with a remembered past and an intended future which characterizes self-consciousness, as opposed to mere consciousness.

We can also, if we wish, think of a conscious, but not self-conscious animal as having beliefs about the world, but we have to be careful to understand just what might be meant by claiming this. What crucially is meant—and what is the sum total of what the evidence entitles us to say—is that the animal has certain dispositions to respond to stimuli in particular ways. What we cannot say is that the animal is aware of its beliefs, or (what comes to the same thing) is aware of itself as having beliefs. (It comes to the same thing because to be aware that I have a *belief* is, looked at from another angle, to be aware that someone— namely, *I*—has that belief.) To say that the animal is aware of its beliefs is to say that it is not just conscious, but also, to some degree, self-conscious.

What I am suggesting here is that the widely criticized behaviouristic account of belief, whereby beliefs are just dispositions to act in certain ways, is in fact the truth about animal belief (or, at least so as to pre-empt certain questions about self-conscious animals, true about the beliefs of those animals which are not self-conscious). In other words, non-self-conscious animal consciousness is first and foremost consciousness of stimuli. It is not consciousness of the plans and con-

---

[5] M. Warnock, *Imagination and Time* (Blackwell, Oxford, 1994), 113.

cepts by which one makes one's way around the world, which in the terminology currently favoured by philosophical theorists could be described as maps or representations of the world. In order to become conscious of oneself as being guided by a representation of the world, one would have to become conscious of what that representation stated. The representation would, in other words, have to be present to one's mind as an objective phenomenon. It would have to stand before one in some symbolic form, as an object apart from oneself and apart from the world, an object which one understood as representing the world in a particular way. And this need for an objective symbolic representation of the world is something which is hard to see being met by dumb animals, or indeed by any creatures we cannot see as possessing the conceptual resources of a symbolic language. Without the possession of symbolic modes of expression, an animal can hardly represent to itself a set of ideas about, but at the same time separate from, the world: it can react to the world, feel the world, enjoy the world, be hurt by the world, but it cannot, in a significant sense, think about the world; it cannot have thoughts about the world. It cannot, therefore, have beliefs which it is conscious of having, however much we might find it convenient to analyse its behaviour in terms of dispositions to act on such-and-such a belief, and however much complicated neural wiring links the stimuli it feels to the response it makes to those stimuli. Talk of belief in the animal case is at most a shorthand for talk of behaviour and of dispositions to behave.

The problem of belief ascription in the case of non-linguistic animals is familiar enough. A dog, we are told, believes its master is at the door. But it can be said to believe this only in a *de re* sense. It believes, of its master, and of the door, that one is at the other, but whether the dog has the concepts of master or door is highly dubious. Which concepts does it have to fill the spaces we fill with the notions of master and door? Again there is no way of telling. As far as we can see, the dog has no symbol system or conceptual scheme in which the master and the door are referred to in a particular way, implying that these things stand in specific relationships to other things dealt with by the system. In the absence of any such system, it makes no sense to ask questions about the concepts under which it refers to entities in the world. It would be methodologically far more economical and far less risky to cut the Gordian knot, and to say that when the dog jumps up at the door on hearing a certain footfall outside, it is simply responding to certain well-known and well-loved stimuli. The relationship between the dog and the man and his gait is a causal relationship, one of stimulus evoking response, and not one of meaning or of (Fregean) sense. The latter, but not the former relationship requires on the part of the possessor a system of mutually sustaining and interlocking symbols by which reference to things in the world is secured via particular meanings.

We see here two fundamental and connected properties of belief, namely intentionality and holism. Intentionality is failure of substitutivity in belief-contexts. Johnny believes that his master is an honest man, but he does not thereby believe that the robber of the High Street bank is an honest man, even though his master is the robber of the High Street bank, because Johnny neither knows nor believes this. At the same time, thinking of someone as his master implies that Johnny has a grasp of the conventions and institutions involved in the relationship of master to pupil, and is able to draw the appropriate inferences when he thinks of someone as being a master and someone else as his pupil. Precisely because having beliefs about things or people involves thinking about them and hence conceptualizing them in specific ways, beliefs never come in single atomic units, but always in clusters, in which one belief implies or presupposes a host of others, which are brought along by virtue of the concepts deployed in the original one. (This is true to a small, but not insignificant way even in the so-called protocol statements of the logical positivists: 'Green here' brings with it not just a notion of the particular colour quality one is currently experiencing, but also, and crucially, the scheme of colour classification in which green occupies a specific position in relation to the other colours in the scheme.)

As I am speaking of the believer's conceptual scheme, it is an abstract system of meanings, by which things in the world are symbolized in particular ways. In the Saussurian distinction, this is *langue* rather than *parole*. Naturally we think of *langue* as an abstract system: its messages can be expressed in various ways; they can be written or spoken or put into some sort of computational system. But, in practice, a symbol system, or at least its messages, are always expressed in some form or other. The imponderable questions about the meanings which might be attached to the beliefs of animals arise precisely because the animal appears to possess no actual form, such as words or writing, by which it might express its meanings. A human child, by contrast, begins to learn a language, and begins to enter a world of meanings which mediates between itself and the world. Through the objective expression of linguistic utterances it begins to approach the world through the specific meanings which language directs to things, rather than simply through the causal effects things have on it.

We can now begin to see why sceptical reflection about one's beliefs presupposes that one is already part of a language-using community and why in a pragmatic way such reflection refutes scepticism about the external world, and has already taken the inquirer one step beyond a solipsistic position. The forming of beliefs which are self-consciously recognized as such requires that one is already a user of a language or a symbol system, something whose actual existence requires tokens in a physical form, written or spoken as the case may be. In the terms of the old Latin tag 'Nomina si nescit, et cognitio rerum periit': without knowledge of names, knowledge of things will be lost. To fix and recall items from the flux of experience, to be able to use this recalled knowledge in inferences and association, we need words to stand as objective, recallable proxies for the things we want to argue and think about. We also, as Daniel Dennett has argued,[6] need to be able to recall and manipulate what we have learned about the things in thought at will, and for this too something standing for the things is necessary (and it is surely significant that while chimpanzees can, in a sense, recognize sameness-in-difference (when confronted by examples), they cannot do simple reasoning tests about sameness and differences in the absence of labels for 'same' and 'different').

It is through language that we can think about absent things, about generalities and about possibility and impossibility. Through language, we can construct abstract arguments, and justify and criticize our beliefs.[7] Language, therefore, is required for self-conscious belief, because the believer needs to become aware of his belief as something separate from the world it is representing, and hence as his representation of the world. And in order to become aware of some thought of his, he needs also to inspect that thought, become aware of its implications and check its correctness by comparison with what it is a thought of. For all of this, the thought needs to take on a quasi-objective existence, as something the subject can inspect and assess; it cannot exist solely as a set of passing sensations and behavioural reactions. So it must exist as something symbolically encoded, something the believer can retain and think about. For the child, none of this is possible until it acquires language. It is only when it masters language that it can properly be said to have thoughts about things, as opposed to reactions and sensations. I believe that this claim is borne out not

---

[6] D. Dennett, 'Learning and Labeling', *Mind and Language*, 8/4 (1993), 540–8.

[7] The significance of language use, and its relevance to self-consciousness is neatly summarized in ch. 5 of Roger Scruton's *An Intelligent Person's Guide to Philosophy* (Duckworth, London, 1996).

only by observation of children, but also by the impossibility one feels
of holding a thought in one's mind which is not encoded in linguistic
form, or some other form which for the moment substitutes for language. It is for this reason that one is driven to follow Wittgenstein in
referring to Mr Ballard-type reports (of pre-linguistic 'thoughts') as 'a
queer kind of memory phenomenon': not something anyway which a
philosophy of mind is to be based on.

We can refer once more to the distinction made by Gerald Edelman
and, following him, Mary Warnock, between primary consciousness
and higher-order consciousness. As we have seen, primary consciousness is certainly possessed by mature human beings, and also by animals and pre-linguistic children. It is what is involved in basic reactive
sensations and the feelings of pleasure and pain we and other animals
possess, and, according to Edelman, it involves the association of
value-impregnated memories with present perceptual experiences and
categorizations. Rather in the manner of the scholastics' *sensus com-
munis*, parallel stimuli received from the various senses are linked into
a unified scene, which is itself given salience by meanings and values
derived from the animal's past experiences. But all this happens unselfconsciously. In primary consciousness, what is not brought into explicit awareness are the past and the future of the animal or child (or
brain-damaged adult) who is subject to primary consciousness, but
without higher consciousness.

What is involved in the transition from primary to higher consciousness is that the subject of the consciousness does not just have
experiences, but is able, over and above that, to refine, alter, and report
its experiences. For this, at least two things are needed. First, the physical infrastructure in the animal needed to produce a symbol system,
and secondly the existence of a society of similarly endowed creatures
to develop, reinforce, and moderate the symbolic realm.

In our species the underlying physical structure is the presence in us
of an anatomical basis for the production of speech: the larynx,
tongue, teeth, lips, and epiglottis, which are unique to humans and
which make possible the production of articulated and rule-governed
sounds. With this physical basis, together with the developed conceptual and memory centres of the brain, the syntax and semantics essential for a symbolic system can then arise. Categorized objects and
experiences can be connected to produced sounds—and the sounds
organized so as to produce assertions and other speech acts.

Edelman points out that chimpanzees, unlike humans, do not have
the brain basis for the complex sequencing of articulated sounds; so

although in some sense they appear to have concepts and thoughts, they have no true language or speech.[8] What Edelman calls true speech, syntactically articulated, allows for the familiar Chomskyan production of a potentially infinite number of sentences from a finite number of words. It also allows for the symbolic categorization both of experiences and of objects in the world, a categorization which does not lead merely to the having of the relevant experiences and sensations in the immediate present, but which, by being separate from the experiences themselves, permits the subject to articulate and then to reflect on what it experiences and perceives without being engulfed in the immediacy of the experience or perception. This symbolic articulation of experiences and of what is being perceived, and the distancing from immediate experience which is involved, is also the foundation in the subject that he or she is a subject, with a past and a future. In explicitly and symbolically categorizing my present experience or perception, I am *eo ipso* comparing it with the past events to which I have attached the same term or symbolic categorization. And on the basis of my sense of myself as having had a past and as existing in a present, I can reflect on what I know of past and present to anticipate and plan for the future.

One thing which is important to realize at the outset is that, whatever may be true at the level of physics or chemistry or neurophysiology, in the macroscopic world neither external objects nor internal feelings come packaged in predetermined categories. This means that the symbolic categorization through language is not just a reading-off from what is already there. Language plays an active role in shaping our picture of the world and of ourselves and is, in turn, influenced by our interests, feelings, and affections.

While admitting the significant role played by symbolically expressed classifications in our sense that we are selves making our way through a world, stable and independent of us, it might still be doubted that we have yet shown that language and our sense of self are necessarily dependent on social interaction. That they are in practice may be true, but this does not yet answer the solipsist who argues that there is nothing necessary here, and that all the appearance we have of being in a public world may be a systematic illusion.

Indeed, we have been led to introduce no notion of community in elaborating a sense of primary consciousness. All that is involved in that is determination by internal criteria of patterns among the

---

[8] Cf. Edelman, *Bright Air, Brilliant Fire*, 130.

multiple parallel signals received by an animal from its environment. Thus, to revert to the previously cited example of Warnock's, a horse may be afraid to enter a horse-box initially because of some inherited fear of dark and confined spaces. But because of some particular bad experiences of trailers had by this particular horse, this particular horse is particularly difficult to box. Munchie is an individual and reacts in the way he does because of his particular experiences, the plasticity of his mind, and the saliences his experiences accordingly set up in him. But though horses are herd animals, there is no need to introduce any notion of the herd or of a community in accounting for our difficulties with him. Why, then, should we feel compelled to associate higher consciousness with a community of symbolizers?

In part the answer must rest on what we have already observed of the non-natural nature of symbolization. Neither our linguistic characterization of our own emotional states and internal experiences nor that of objects in the world mark out natural kinds existing and demarcated independently of our linguistic systems. Even the characterizations of phenomena as broad and as apparently homogeneous as pain and pleasure involve considerable degrees of open-endedness and indeterminacy: are all pains, for example, unpleasant? Is pleasure something one always wants? Are pains and pleasures mutually exclusive? We do not, I think, need to be students of de Sade or Masoch to see that these questions do not admit of straightforwardly physiological answers, and that there is a degree of social and conventional understanding involved in determining what is to count as pain and pleasure respectively.

It might be said that similar indeterminacies will attend the categorizations involved in primary consciousness, which *ex hypothesi* is neither linguistically mediated nor socially involving. It is certainly true that what an animal finds salient in its present experience will be to some degree idiosyncratic, depending on just what its past experience has been. But this will simply be a matter of causation. That Munchie refuses to go into his trailer may well be due to unpleasant experiences he has had in the past. If the new trailer is perfectly safe and very different from the old one in which he was bashed about, or even if his trailer-phobia was bought about by bad experiences in a narrow loose-box, rather than a trailer, there is no sense that his present stroppiness is a *mistake*. It just exists, and it has the causal antecedents it does; it may be more or less idiosyncratic. In a sense, then, in primary consciousness the justification of present salience—of the way the beam of consciousness shines in a specific case—does not arise. It shines the way it does, and that can be more or less useful, but its being or not being useful is not a matter of its being right or wrong. There is no *norm* against which the beam's current shining is being compared.

It is quite different with symbolic characterizations. There, if I categorize my present sensation as painful or the object before me as a horse, whether my present use of those terms conforms to my past use makes all the difference in the world. Accounting for the causal genesis of a present mistaken use of mine does not stop it being mistaken. In using a symbol to refer to or describe something, I am *ipso facto* entering an arena where mistakes are possible, and there normativity is essential. And its being mistaken matters because using a given term incorrectly will render unsound those inferences and conclusions one would draw from the incorrect classification. Even more, if one is unable to secure consistency in the meaning and use of one's terms, the symbol system will fail practically and theoretically: one will have but a simulacrum of categorization, rather than the real thing, and in its inefficacy and incoherence, the system itself will quickly lapse into disuse. In other words, symbol systems can be effective in firming up and conveying meaning only if they are operated according to the rules governing the meanings of terms and the inferential relationships held to exist between them.

Symbol systems, such as human language, require a community of speakers. There is, of course, empirical evidence for this, of a negative sort, in that unsocialized children do not develop language, and, indeed seem to be unable to develop it if they have not been in human communities at the crucial developmental stages. It is, moreover, perfectly clear that the normal method of language acquisition (in a human community and through interaction with speakers) is going to be far more efficient than inventing a language from scratch. The socialized language-learner will be able to draw on all the resources and concepts of the existing language, and will be guided in his or her learning by other users of the language.

But there are powerful reasons, deriving partly from Wittgenstein's analysis of private language and rule-following, for thinking that a linguistic community is necessary for the production and maintenance of a language, and not just highly desirable.

The point which needs to be emphasized is that there are different possible ways of applying any given concept or rule in the future. Past uses and instructions do not uniquely determine what is to count as a correct future usage or application: this is the lesson to be derived both from the Wittgensteinian analysis of rule-following and from

Goodman's new riddle of induction.[9] In Wittgenstein's analysis, any-thing I do can, on some interpretation, be made to accord with a rule, even continuing the series '+ 2' by saying 1004, 1008, 1012 after 996, 998, 1000. Past instructions will not have ruled out the possibility that after 1000, '+ 2' is to be taken to mean what we would naturally regard as being + 4. Or, if they had explicitly done so, there would be other deviant possibilities they would not have ruled out. The 'we' here is crucial in developing and sustaining a notion of deviance. It is because we are members of a community on the whole united in our reactions to the deviant and the rule-according that there is a genuine distinction to be drawn between the correct and the incorrect, and that what is right is not equivalent to whatever seems to me to be right—a state of affairs tantamount to having no sense of rightness at all.

Goodman's paradox involves the possibility that a past regularity should, in the future, continue in an unexpected way. Thus, we expect emeralds to be green after the year 2000, but there could be a world in which post-2000 emeralds are blue. In such a world, which up to now would be identical to ours, emeralds might be described as grue (= green up to 2000, and thereafter blue). This would be a regularity, and one consistent with what we have so far observed, though for us (green/blue speakers) a strange one. But that just underlines Goodman's point: that regularities or similarities are 'where you find them, and you can find them anywhere'.

The green-grue example is contrived, unnecessarily so, for it is per-fectly possible to present Goodman's point in an uncontrived way. Prior to Einstein, and observation of systems moving with very high relative velocities, all dynamical observations would seem to have con-firmed Newtonian physics. Of course, these observations have not been discarded in the new perspective. It is just that they are general-ized from in a different way, so as to accommodate and predict the dif-ferent types of result at high velocity.

We can reformulate Goodman's point so as to make it apply not to straightforward inductions on past evidence, but to our use of terms and the rules governing their application. Let us call some plant 'T' if it is the same or similar to the plant(s) we previously called 'T'. But what is to count as being the same or similar in the earlier Ts? Similar in what respect(s)? The same in what way(s)? We can, of course, rein-force our intuitions with explicit instructions, but the instructions

    [9] Cf. L. Wittgenstein, *Philosphical Investigations* (Blackwell, Oxford, 1953), pt. 1, sects. 198 ff.; N. Goodman, 'The New Riddle of Induction', in his *Problems and Projects* (Hackett, Indianapolis, 1972), 371–88.

cannot exclude every possible deviation, and in any case the terms in which the instructions are framed will themselves be subject to various vaguenesses and indeterminacies.

From my own point of view, in trying to establish correctness in the use of my terms, I will have only my own reactions to go on. At a certain fundamental level, I will reach bedrock in my judgements, and there will be no difference between what is right and what seems to me to be right. And the same will go for every other user of the language, individually considered. Particularly given that terms are in many ways vague, open-ended, and very far from simply mirroring either the laws or the kinds of nature, the only way out of the impasse would seem to be for some group correction and development of language. What seems in general to be correct introduces an element of negotiation, cross-checking, and distance for immediate and particularized usage which is not present when there is only my current reaction and memory to go on.

If, then, language is inherently social (for its stability) and if self-consciousness requires a symbolic system such as language for its realization, it follows that I am conscious of myself only in so far as I am a member of a community of agents, who are using the language, and who similarly conceive themselves to be self-conscious. There is a hint of circularity here. What is being said is that self-consciousness and language hang together and that language is a social phenomenon. So I can only be self-conscious, if I am in a group of people who also speak and see themselves as self-conscious. So given that all the members of my community also require the rest to become self-conscious how does anyone speak and become self-conscious in the first place?

Infants accompany their activity with sounds, 'scribble talk'. After a while scribble talk matures into real speech, as it latches onto the already existing language that surrounds them. Mere signals or random responses become concepts with standards of correctness in their use. But how did the process get started in the first place, from ur-speakers, grunting and signalling automatically without thought or intention, to speakers of languages in the full sense? The answer to the question is shrouded in the mists of time, and we can only gesture at the answer: a community of proto-humans gradually gets into the state where they can (suddenly?) act as language users. The difficulty of answering the question, though, does not impugn the steps by which the question arose, for those steps were forced on us by reflection on the nature of self-consciousness and of language.

Given, though, that language is social, language also allows us to enter into new kinds of social relationships with others, ones in which we can give precise expression to our emotions and also ones in which we can criticize and judge our own behaviour and that of others.[10] The importance of this aspect of our linguistic inheritance for social life—that which makes morality and politics possible will concern us further in Chapter 6.

In what I have just argued, I have travelled a Wittgensteinian route to the essential sociality of language. To reinforce the point, and to bring out some other aspects of the issue, it is worth glancing briefly at some observations of C. S. Peirce on the intersubjectivity involved in coming to self-consciousness. As will become clear, in his remarks Peirce underlines the intersubjectivity of both language and self-consciousness.

We may begin, as Peirce suggests, in thinking of a child acquiring a language. Peirce prescinds from the question as to whether the language is public or private. What he wants to show is that a systematic connection in a child's mind between certain sounds and certain facts is not yet enough to bring about self-conscious belief on the part of the child. For this he needs the experience of testimony and speech which does not simply reinforce his own grasp of language, but which actually contradicts his own experience. It is, in other words, the knowledge that the testimony of others can correct the child's own beliefs which forces on the child's attention the fact that he himself is a believer, a creature, that is, who symbolically represents the world to himself in a way which might be wrong, and of himself as the locus of this potentially false representation.

In his paper 'Questions Concerning Certain Faculties Claimed for Man', Peirce argues that when a child begins to converse, he begins to learn that what other people say is the best evidence of fact which can be found, 'so much so, that testimony is an even stronger mark of fact than the facts themselves, or rather than what must now be thought of as the appearance themselves'.[11] The suggestion is that through participation in public discourse the child first, and then the adult, comes to see himself as a member of a community of inquirers, in which no individual has a privileged position, but in which everyone's experiences are subject to the judgement and corroboration of the rest of the

---

[10] Cf. Scruton, loc. cit.
[11] C. S. Peirce, 'Questions Concerning Certain Faculties Claimed for Man', in *Charles S. Peirce: Selected Writings*, ed. Philip Wiener (Dover Publications, New York, 1958), 28.

community. 'I may remark, by the way, that this remains so through life; testimony will convince a man that he himself is mad.'

Through participation in a speech community, the individual comes to realize that when sometimes the testimony of others is contradicted by his own current idea it is testimony, not his idea, which is correct. In this realization, the child becomes at the same time aware of his own ignorance and of himself as the subject in which this ignorance inheres:

A child hears it said that the stove is hot. But it is not, he says . . . but he touches it and finds the testimony confirmed in a striking way. Thus, he becomes aware of ignorance, and it is necessary to suppose a self in which this ignorance can inhere. So testimony gives a first drawing to self-consciousness.[12]

So, far from the Cartesian model in which self-consciousness is a prime datum underlying all other knowledge, Peirce's view is that self-consciousness is something which arises only when one sees oneself as a fallible and at times dissentient member of an already existing community of speakers and inquirers. Even though once we are mentally mature we might be more certain of our own existence than of any other fact, this is not because our own existence cannot have been inferred from any other fact. It is, in Peirce's view, inferred from *every other* fact of which I am aware, but without the existence of and certainty of at least some of these other facts, I would never have been in a position to infer my own existence, to regard it as something of which I am explicitly aware.

The Peircian view then is that self-consciousness arises in and through my awareness of the fact that I am a believer. Self-consciousness is initially consciousness of myself as a believer, which in turn depends on my being confronted with the exosomatically (linguistically) expressed beliefs of others, which stand in complex relations of harmony and discord to my own experience and ideas. In seeing this, I become aware that I too am making my epistemic way through the world, which is also populated by other believers who are able to lay down epistemic and conceptual norms. I thus come to formulate and to create my own beliefs. (I say that I create my own beliefs: for although I might have had unconscious dispositions to react to the world in certain ways prior to belief formulation, as our reflections on intentionality suggest, it is only in and through the formulation of dispositional beliefs that their precise meaning and sense becomes clear.)

---

[12] Peirce, loc. cit.

So self-consciousness and full-blooded belief depend on two things: membership of a community of testifiers and disagreement with and later correction by at least some of that testimony, leading to one's sense of oneself as a focus of ignorance and, by the same token, of belief and at the same time a sense of oneself as entering into the conceptual system of those around one. No doubt the child experiences loss and frustration at times, as well as satisfaction and unity at other times in his or her pre-linguistic experience, as presumably do members of other species. Its habits sometimes succeed and sometimes fail, and these successes and failures evoke feelings of pleasure and pain, fulfilment and satisfaction. But none of this habit and instinct-based interaction with the world amounts to the conscious entertaining of beliefs about the world. For that the child has to have access to and awareness of a system which stands between it and the world, which enables it to represent the world to itself, and so become aware of itself as a believer. The Peircian thesis is that it is through being confronted with the linguistic testimony of others conflicting with its own experience that the child first comes to realize that there is a system of representation into which he is being inducted and in which competing versions of the world can be expressed. The child has access to the system; in becoming aware of this and in beginning to use the system the child in turn becomes conscious of itself as a believer and as a self. It is no longer an organism reacting to the world. It begins to see itself as a centre of consciousness making its way through the world, with its own beliefs and ideas as to how the world is. But it does so only in contrast to the others around it who also have and express beliefs about the world, and indeed about itself. So, in seeing myself as conscious, in being self-conscious, I am already situated in a world with other self-conscious believers and speakers also expressing their ideas about the world, and about me.

Naturally in having this idea of himself, the self-conscious speaker and believer is aware that his use of language and his beliefs might be wrong. In becoming aware that one is a believer through becoming aware of one's own fallibility, the very conception of oneself as a believer brings with it a hint of scepticism. One learns that one's own beliefs and descriptions do not always fit with the world. Perhaps they might be very far from fitting; perhaps they might be wrong in quite significant ways, not about just whether the stove is hot, but about the very nature of the stove. Perhaps it is not at all as it seems. Perhaps it is not solid or coloured or hot in reality; perhaps all this is just due to my perceptual constitution and not to the nature of the stove itself.

With this type of radical questioning of one's own beliefs it is but a short step to the conclusion that one's original relationship to the world is like that of a blind man fumbling around a room whose appearances entirely escape him, and which he can only infer from the hints and suggestions afforded by the sense of touch. Of course, the epistemic situation is worse than that: touch and all the other senses are just as likely to be wrong as vision. On this view, we know nothing directly about the external world: if it exists at all, its true nature forever escapes us. Scepticism this extreme—that there is an external world, but it might be utterly different from our perception of it—comes close to solipsism, because the world I experience *is* my creation.

But this extreme form of scepticism, tending towards solipsism, is a case of the reflectively self-conscious believer forgetting the way his reflectiveness is based in his initial grounding in the world and in his participation in a community of speakers. A very real question must therefore be raised about the extent to which an inquirer can legitimately cut himself off from the starting-point which made his inquiries possible in the first place. If, as I have argued, reflective inquiry and language can take place only given that one is already in a public world populated with other inquirers, and responding to the world and to what other inquirers say, there is clearly something problematic in extreme sceptics using that form of inquiry to cast doubt on the nature of the world as revealed in the assumptions of the common linguistic and conceptual framework. One would, at the least, appear to be in the position of mariners replacing all the planks of their ship at once. We shall, in subsequent chapters, examine in more detail the implications of this doctrine for both epistemology and ethics.

Nevertheless, it is important that we do not go to the other extreme and attempt to deny that reflection and reflective questioning have any genuine role to play in our lives and practices. That they do is part and parcel of the fact that we do not simply respond to stimuli, that, by virtue of our nature as speakers and self-conscious agents, our responses, verbal and otherwise, to the circumstances in which we find ourselves are guided by a sense of what we find appropriate or fitting in a given situation. Part of the meaningfulness of what we say and do arises from the fact that our behaviour, linguistic and otherwise, is not blind. It is always guided by our understanding the situation in a certain way, under a certain conception. We can, then, defend what we do by appeal to this conception, which suggests both that the possibility of reflecting on what we do and of alternative responses is a crucial aspect of human as opposed to animal behaviour. It will be important to remember the key place of reflectiveness in human life when we come to consider conservative theorists of ethics who regard the existence of self-consciousness about what we do as inevitably a sign of the decadence of a culture.

Nor are we saying that because being part of a linguistic community is the condition of self-consciousness and reflectiveness, we are thereby bound to agree to any particular beliefs or values, however widely shared they may be in our community. It is precisely because we are members of a community that we become self-conscious and *thereby* each have an obligation in reason at least to take responsibility for the beliefs and values we adopt. This is not a paradox, but a conclusion forced on us by thinking about self-consciousness and language. It seems highly likely that only creatures with a specific mental (or even spiritual) make-up are potentially self-conscious. It is striking that in our world creatures other than human beings do not uncontroversially manifest signs of self-consciousness or indeed of language. The experiments with chimpanzees, such as Washoe, clearly involve— from the chimpanzee perspective—highly unnatural arrangements (and they do not in any case convince all involved that language mastery has occurred). Even if a few apes can after considerable training be inserted into linguistic communities, speech for them is neither natural or normal. By contrast every human community is a linguistic community, however different other circumstances are. So our being self-conscious and also users of language tells us something about what we are and about what our natural potentialities are. But any potentiality, however deeply rooted in a creature's nature, needs the right circumstances to manifest itself. It remains true, though, that it is only within a linguistic community of similarly endowed beings that our potentialities for self-consciousness are realizable.

At the same time, it may well be right, as Durkheim (among others) has noticed, that some types of social arrangement are more conducive to the growth of the reflective spirit. Durkheim says that there are societies in which social solidarity is so highly developed that the individual human being does not appear. In such societies everything the individual does or thinks follows the norms and beliefs of his society. Durkheim contrasts this type of society with that in which each individual has, as he puts it, a sphere of action which is peculiar to him. He sees the development of the individual personality going along with increased specialization, with the division of labour. A society with a great deal of specialization and highly differentiated labour will, accordingly, produce many more highly differentiated individuals than one in which everyone simply follows the collective norm and will.

While there may well be something in the claim that reflectiveness and self-consciousness are more evident and more pronounced in some societies than in others, and while this phenomenon may coincide with societies in which there is considerable division of labour, I do not believe that the two are necessarily connected, nor that there is any human society in which reflectiveness is altogether lacking.

The reason for this is that, as we have just seen, reflectiveness is involved whenever human beings speak and act in a meaningful way, as opposed to reacting like animals. Even if a society discourages questioning of basic society-wide assumptions or principles, as some societies certainly do, every individual within the society will be encountering new situations all through his life. He cannot simply mechanically apply what has been said or done in the past. Reflection, thought, self-conscious speculation will be required even to decide how what was said or done in the past is to be said or done now. In this sense, every society in which people speak or act on rules and in the light of beliefs contains the seeds of its evolution, because its members engage in criticism of its practices precisely in adapting its practices to new circumstances. This criticism need not be criticism from an external point of view, so to speak, trying to judge its practices by reference to standards or goals other than those already implicit within the practices. Sometimes when conservatives and others deny that life should be led by reference to rationalistic reflection rather than by unthinking habitual behaviour, what they are attacking is a certain type of rationalism, not that distant from scepticism, which would judge everything by reference to aims and standards appropriate to other activities than the one in question. But they are wrong to think that any human activity is unthinking or unreflective. The presence of thought, reflection, and self-conscious belief is what makes human activity different from the conscious but unreflective behaviour of non-linguistic animals.

4

Evolutionary Epistemology

We must presume that human beings are products of evolutionary development. Given this, what, if anything, follows about our knowledge? Can any epistemological conclusions be drawn from the fact of evolution? If so, what might they be? In exploring these questions, we will highlight the strengths and weaknesses of evolutionary accounts of human activity more generally. In general we will be drawn to conclude that while our activities, including our knowledge-gathering activities, are rooted in our biological inheritance, as human beings, operating in a human-cultural world, we have taken on goals and activities whose aims and rationalia cannot be explained in biological terms. In many ways, the examination of knowledge is a keystone here, for there is clearly something biologically advantageous in having knowledge, or at least, in having survival-promoting beliefs. As we shall see, though, while we can learn some interesting things from evolutionary theory about survival-promoting beliefs, matters become less clear when we look at knowledge in the full sense.

In approaching this question, the first thing to note is that while Darwinism explains natural phenomena broadly in terms of adaptive utility, within this general framework it allows that an organ or faculty may persist in a species so long as it does not worsen the survival and reproductive chances of members of that species relative to its competitors. Thus Darwinism will allow the persistence of organs and faculties which are not in themselves useful in generally well-adapted species, so long as they do not detract from the adaptiveness of its members as a whole. In this chapter, I will suggest that human cognitive powers do undoubtedly have some useful or adaptive functions, but that they also have other aspects which have nothing to do with survival, and which may even prove detrimental to survival. I will argue that our truth-seeking drive falls into this category. Seeking the truth in inquiry is a different aim from that of providing oneself with beliefs which are useful for action. The truth-seeking aim is linked logically with the possession by us of self-consciousness, and hence I will also have something to say about the adaptiveness of self-consciousness.

In considering non-adaptive aspects of our cognitive drives, and of self-consciousness itself in so far as it grounds these drives, we shall be putting some flesh on the bones of the oft-repeated claim that in human knowledge we get something altogether new and un-Darwinian in the natural world. This may seem to raise questions as to where this new thing comes from, and as to the adequacy of seeing it as a merely non-dysfunctional adjunct of powers which are functional.

To some, indeed, the significance has seemed to cast doubt on the adequacy of Darwinian explanations of human powers, a point to which I shall return later in the chapter.[1]

Life and Knowledge: Nine Supposed Parallels

A number of features of biological evolution have been seen by those who would class themselves as evolutionary epistemologists as relevant to knowledge and human culture generally.[2] These features suggest parallels between human knowledge and evolutionary processes in terms of both aims and structures. They are:

1. Life is distinguished from inert matter in that an organism has to do something in order to remain in existence to survive and reproduce. This 'doing something' at the crudest level involves the ingesting of food from its environment.

2. It can be regarded as having to solve problems thrown up by the environment, to maintain itself in equilibrium both internally and in relation to its environment.

3. Living things, in order to draw sustenance from their environment, tend to create scarcity in solving their problems.

4. Living things address themselves to their environment initially through preformed dispositions and expectations.

5. The evolution of species may be seen in terms of developing long-term solutions to short-term problems—that is, developing better devices and dispositions to cope with and feed off the environment.

---

[1] See particularly Thomas Nagel's *The View from Nowhere* (Oxford University Press, New York, 1986), 78–82.

[2] In this analysis I follow the summary of the claims of evolutionary epistemology given in Gerard Radnitzky, 'An Economic Theory of the Rise of Civilisation and its Policy Implications: Hayek's Account Generalised', in *Ordo (Jahrbuch für die Ordnung von Wirtschaft und Gesellschaft)* (Gustav Fischer Verlag, Stuttgart), 38 (1987), 47–89, at 50–2.

6. The mechanism of evolutionary development is that of blind variation and selective (i.e. environmentally selected) retention.

7. Long-term solving of short-term problems has led most crucially to the development of cognitive sensory apparatuses. This reduces risk to life, vision and the other senses replacing collision, and leads to great economy of locomotive effort.

8. The crucial point to note is that sensory apparatuses are seen by evolutionary epistemologists as evolving from a quest for food, rather than from a quest for information or knowledge for its own sake. Thus according to Gunther Wächterhäuser, the prototype eye in primitive marine organisms may be seen as having evolved because, for the bearers of this eye, light was a source of food: 'The type of light searched by photomovement was automatically "edible" light useful for photosynthesis.'[3]

From the biological and evolutionary perspective of these eight points we can see that perceptual and cognitive faculties develop so as to enable creatures to cope more successfully with the problems thrown up by their environment, thus to survive the better and, as a consequence, to reproduce the more. This may seem to lead to some suitably optimistic epistemological conclusion, about our faculties representing the world accurately because they have developed just so as to do that.

9. Finally, it is also often claimed (by e.g. Popper and his followers) that these faculties themselves work on the same basis as other evolutionary processes, that is, by means of what Popper is prepared to refer to as *blind* variation and selective retention. That is to say, the perceiver approaches his environment with a set of assumptions about what he is seeking, but has yet to find.[4] The vision of a frog, for example, is seen as an in-built disposition to respond only to certain stimuli (moving objects in the frog's environment), as a result of which the frog is enabled to react. This process is regarded as blind because the frog's eye is already conditioned to perceive and react only on certain stimuli and is blind to the rest:[5] its vision itself is 'theory-laden', and is not to be seen as directly mirroring the environment. Selective retention comes in because sometimes the visually inspired 'theory' that

---

[3] In 'Light and Life: On the Nutritional Origins of Sensory Perception', in G. Radnitzky and W. W. Bartley III (eds.), *Evolutionary Epistemology: Theory of Rationality and the Sociology of Knowledge* (Open Court, La Salle, Ill., 1987), 121–38, at 124.

[4] Cf. K. R. Popper, 'Replies to my Critics', in P. Schlipp (ed.), *The Philosophy of Karl Popper* (Open Court, La Salle, Ill., 1974), ii. 1061.

[5] Cf. K. R. Popper, *Objective Knowledge* (Clarendon Press, Oxford, 1972), 145.

there is food in such-and-such a direction is confirmed (by the frog's tongue trapping a fly) and sometimes it is refuted (by the frog's failure to get any food or to avoid a predator). Selective retention is also operative at the level of organs and individuals. Individuals with sensory faculties good for solving their short-term problems will tend to survive and reproduce more than their competitors, and so their sensory faculties, too, will be retained.

While at a level of high generality there are certainly some similarities between biological evolution and the nature of human knowledge, one can also point to significant dissimilarities at a more focused level of comparison on at least eight of our nine points. These differences suggest that there can be a unified theory concerning biological evolution and human cultural evolution (as Radnitzky, for example, urges) only at the most general level.

1. A knower or perceiver in knowing or perceiving does not literally take anything from the environment, even if his knowledge or perception later allows him to do this. (The question as to what, if anything, a knower or perceiver physically takes in from the environment dates back at least to Aristotle.)

2. It is by no means clear that all knowledge and perception is a response to a problem thrown up by the environment, at least not if problems are construed in terms of the survival of the individual knower or perceiver or the species to which he belongs. (We shall return to this point in considering Dewey's claim that conscious perceptions and ideas always result from some disturbance of settled habit, some disequilibrium between organism and environment.)

3. As knowledge and perception do not result from one's literally taking anything from the environment, it is not clear that the scarcity point applies directly here. My perceiving Buckingham Palace or knowing about the Royal Family does not make it more difficult for you to do these things.

4. There is a parallel which could be urged here. Knowers and perceivers do come to any situation with a battery of expectations and dispositions, although, as we will see, the control of these pre-existing dispositions over what is subsequently perceived or believed can be very loose and plastic, in contrast to the so-called tropistic behaviour of organisms and animals who can respond to circumstances only according to preformed patterns of response, and who manifest no flexibility in response to those circumstances. Similarly, it is argued that the immune system is activated just when an antigen invading the

system activates a specific and pre-existing antigen-recognizer in the host system with a complementary molecular configuration.[6] But human perception is surely far more flexible than this. We are guided in our perception by pre-existing categories, but not constrained by them rigidly. We can recognize and track objects that do not fit into existing categories, but which instruct our minds to form new categories.

5. While our cognitive faculties can be seen in terms of long-term solutions to short-term problems, I shall argue that they should not be seen solely in these terms: there are also the possibilities of disinterested activity that they bring, aesthetic, moral, and cognitive. One should also be wary of analysing our actual theories and perceptions in terms of long-term solutions to short-term problems.

6 and 9. To regard perception and theory-building in terms of *blind* variation and selective retention overlooks the rational element in theory-building and the 'sighted' sensitivity to our environment of normal vision, hearing, smell, and the rest. On the latter point, 'blind' is surely metaphorical as applied to perception. The point of the metaphor is to suggest that the perceiver is like a blind man fumbling for something he suspects is near him, but is unsure exactly where, and in terms that are already available. Against this, though, surely much perception is precisely not like that, but more a matter of coming to realize that hitherto unsuspected objects are around one, objects sometimes categorized in terms one already has, but by no means always. Perception follows or tracks objects it has not suspected as much as it fumbles for ones it suspects in terms of pre-existing categories. Theorizing is not *per se* blind either. While there may be some intuitive and imaginative leaps into the dark, hypothesis-forming is characteristically guided by and responsive to evidence and changes of evidence. Thinking of human perception and theorizing as if they were random or blind processes, like genetic mutation, and simply determined by pre-existing genetically based forms of thought, grossly underestimates the extent to which they are guided by evidential and rational considerations, and instructed in what we know and perceive by environmental inputs. That there is an element of instruction and guidance from without is even more clearly the case with the development and formulation of scientific theories, contrary to much Popperian rhetoric.

---

[6] Cf. Ian J. Deary, 'Applying Evolutionary Epistemology from Immunity to Intelligence', *Journal of Social Biological Structures*, 11 (1988), 399–408.

As John Worrall has put it, in a recent review of the situation, 'the articulation of new theories is *not* (unsurprisingly) a question of throwing out possible conjectures "at random" and then subjecting them to rigorous selection pressure; the process is not (even approximately) analogous to Darwinian natural selection'.[7] What is far more typical in the history of science in the development of new theories is the entirely rationally reconstructible result of plugging new data and judgements based on data about other possible theories into background knowledge. The process is, thus, far more gradual, rational, and goal-directed than talk of blind variation and selective retention would suggest. Such talk, indeed, is largely misleading as an account of scientific theory construction, as misleading in its own way as the old quasi-Baconian model of the scientist simply reading theories off from masses of undigested observations and in no way more flattering to the rationality of scientific investigators.

7. If we accept the evolutionary story, our cognitive and perceptual faculties have no doubt developed against the background of the search for long-term solutions to short-term problems, and it is not implausible to see them, as Popper and his followers have done, as agents of vicarious locomotion, reducing the risk of collision, and so on. It is not implausible to see them in this way, but we should not overlook the fact that contemporary Darwinian theory does not require that every adaptation or disposition of a surviving species has itself promoted the survival of individuals of that species. So long as it does not significantly reduce the survival chances of its possessor compared to the chances of competitors, a quite useless faculty or disposition may not be weeded out by the environment. Sometimes even potentially pernicious faculties or dispositions may survive in a species until the potentiality is revealed. Perhaps in the case of a whole species a disposition favourable for a time in its development is in the end its undoing. So we should beware of inferring from survival to fit between species and environment, or adaptability, and also of inferring from past survival, even aided by a given faculty, to future survival or to the future helpfulness of that very faculty.

Generalizing the caveat against too optimistic a conclusion from evolution to epistemological justification, it is also worth pointing out that the theory of evolution cannot provide any refutation of global scepticism, or in any way solve Kant's problem of the unknown noumenon. While Popper and his followers—being hostile to talk of

---

[7] John Worrall, 'Revolution in Permanence', in *Karl Popper: Philosophy and Problems*, ed. A. O'Hear (Cambridge University Press, 1995), 75–102.

justification in any sense—would not have argued in this way, Konrad Lorenz certainly did. He suggested that while Kant was correct in focusing his epistemological gaze on our cognitive faculties, he was unable to penetrate the veil he thus interposed between our faculties and the world because he failed to recognize that our faculties were the product of evolutionary interaction over millennia between the world and our ancestors.

As Lorenz puts it:

What a biologist familiar with the facts of evolution would regard as the obvious answer to Kant's question was, at that time, beyond the scope of the greatest of thinkers. The simple answer is that the system of sense organs and nerves that enables living things to survive and orient themselves in the outer world has evolved phylogenetically through confrontation with and adaptation to that form of reality which we experience as phenomenal space.[8]

While this is doubtless correct from a biological point of view, it is hard to see how it can stand as a reply to 'Kant's question', or, indeed, to any form of radical scepticism about the real world or its basic nature. For the truth of Lorenz's observation presupposes the correctness of biological theory. It presupposes that the world is much as the theory of evolution tells us. It is just this knowledge which Kant and radical sceptics are putting into question: couldn't the world be very different from how it appears to us, even in our best theoretical account? There may be reasons for wanting to disallow such scepticism, reasons we will touch on in the next chapter; but ruling scepticism as ill-conceived is not to think that one has any knock-down answer to it, particularly not an answer which does not turn out almost at once to be circular.

In any case, even if the circularity of Lorenz's strategy were overlooked, it is far from clear that it could deliver any telling blow against Kant. Kant, after all, is not denying that there is a real world, nor that the real world appears to us in accordance with empirical laws. What he denies is that its appearance to us in accordance with empirical laws can be shown to be a correct account of how it really is. Now, accepting Lorenz's premiss, it is true that our perception and theories have been useful to us in charting and predicting the future course of our experience. But, as we shall see further in the following sections, it does not follow from that that our perceptions and theories are particularly accurate representations of reality. As Lorenz himself points out, evolution is quite consistent with the (very successful) survival of

[8] Konrad Lorenz, *Behind the Mirror* (Methuen, London, 1977), 9.

creatures who perceive a spatial world of four or more dimensions in terms of one-dimensionality, and a continuous flux of forces and events in terms of stable and enduring material objects. In what sense can biology be held to refute Kant if it is consistent with a real world so different from the way it appears to members of highly successful and long-lasting species? And Lorenz himself says in discussing the one-dimensional representation of the world possessed by the paramecium, even we cannot know how many dimensions there are to space *an sich*. In what sense can any of this be a *refutation* of Kant, or of scepticism more generally?

8. It is here, though, that we come to the crucial point, the one which, if nothing else, demonstrates the strict irrelevance of evolutionary epistemology to epistemology. For even granting what, despite our initial caveats on (7), is probably true—namely, that overall the cognitive faculties of a successful species are useful to it in its struggle to survive and reproduce—it does not follow that what is thereby revealed is either true or, except in a highly restricted sense, accurate. There is a clear distinction to be drawn between the true and the useful, the very distinction indeed which allows us to question the truth of the paramecium's one-dimensional world-view, or even our own three- (or four-)dimensional perspective. And as I will now argue, this distinction is deeply embedded in the structure of our cognitive activity itself, in the shape of a logical demand to search for what is true and reasonable, and which cannot be analysed in terms of the useful.

The True and the Useful

In so far as we are self-conscious agents, we realize that there is a distinction to be drawn between ourselves and the world in which we exist. The very notion of self-consciousness implies that one is conscious of certain experiences as being had by oneself. But this in turn implies a notion of experience; which in turn implies a contrast between how things are experienced and how they are. Self-consciousness involves a conception of oneself as opposed to other things. This consciousness of oneself as separate from the rest of reality then opens up a distinction between how what is not me really is and how it appears to me. In becoming conscious of myself as separate from the world then, I have to entertain the possibility that the world might in various respects be unlike the way it appears to me. My perspective on the world might be partial and distorted in various ways. In seeing myself as separate from the world then, I have to allow that the world has an existence apart from me and to concede the possibility that it may not be as I think it to be. My self-consciousness then, is at once a consciousness of the otherness of what is not me, and of its potential separateness from my view of it. If the world were necessarily as I perceive it to be, it would not be truly other, and so could not form the contrast necessary for my specific self-consciousness.

The realization that the world might not be as I perceive it and believe it to be at once raises a problem for me regarding my beliefs and perceptions, at least to the extent that I am conscious of them. As we learn from Moore's paradox, one cannot simultaneously assert p or assert that one believes p and fail to hold that p is true. There is in the very logic of assertion and belief a directedness towards what is true, towards what is really the case. To waver over the truth of what one believes or asserts is to shift from belief or assertion that p to mere entertaining or supposing that p. In belief and assertion proper one intends a movement from one's subjective view to how the world really is. Belief and assertion in this full-blooded sense are indeed concomitants of one's existence as a self-conscious agent operating in a world one recognizes as separate from one's initial beliefs and assertions. One might go further here, and assert that one's self-consciousness throws up not just two possible worlds: the real world and the world constituted by my beliefs, but many possible worlds corresponding to many possible belief-systems. After all, my second set of beliefs after correction by the real world might still be way off the truth, and this is something a self-conscious agent reflecting on his state must readily admit. And if I see others as perceivers and agents, I must envisage the possibility of their having different belief-systems too and belief-systems which could in principle challenge and correct my own. There is thus plausibility in the suggestion that full-blooded belief in a self-conscious agent implies acceptance of a many-possible-world ontology, or at least of the mental structures that might ground such an ontology.

Once we see that a notion of objective truth is central both to our self-consciousness and to our having beliefs and making assertions, it is easy enough to see that there is a distinction to be drawn between what is true (= what *ought* to be believed) and what it might be *useful* to believe. In everyday experience there are plenty of examples which point to the validity of the distinction, such as the use of Ptolemaic assumptions in navigation and of Newtonian mechanics in everyday matters. While we know that these theories and assumptions are false, using them is far easier and more convenient than basing our calculations on the truth. Indeed, using the truth in these cases might well be so cumbersome and difficult that we might not act quickly enough when we have to make a calculation on which to base action. This aspect of the distinction between the true and the useful has actually been made an important plank of the evolutionary account of our perceptual faculties by Donald Campbell (in his unpublished William James lectures on Descriptive Epistemology and elsewhere). Rather against the optimistic tenor of Lorenz's use of evolutionary consideration in discussing epistemology, what Campbell says harmonizes well with the minatory note struck in the last section.

In his second lecture, Campbell points out that 'for those of us with conscious experience' what he calls 'slight and fuzzy' environmental differences are represented by 'vivid, clear, over-clarified perceptions of reality'. Our sense of taste, for example, transforms very slight nutritional advantages in given foods (of the order of a one per cent improvement in the reproductive and survival chances of those who choose the one food over the other) into an absolute all-or-nothing preference for the one food over the other. Our vision, too, presents objects to us as sharper and more clearly demarcated from their surroundings than according to physics they really are; according to Campbell, 'the vividness and phenomenal directness of vision needs to be corrected in any complete epistemology'.[9] Along similar lines, Quine points out that slight differences from one species to another in the sensory mechanisms for colour categorization result in overwhelming differences in the resulting grouping of things by colour.[10] What from this perspective might be regarded as a visual distortion here has, as Quine suggests, considerable importance at the food-gathering level. Once again, an epistemologically questionable perceptual disposition can be justified in terms of its usefulness for survival and presumably has been selected because of that.

Again, for a creature to jump away from any predator-like shape in its environment may imply all sorts of errors in its beliefs, but being correct here one time out of ten would convey a distinct advantage to the possessor of such a disposition over his fellows who, while correct epistemically nine times out of ten, were wrong on the all-important

---

[9] Donald Campbell, 'Evolutionary Epistemology', in Radnitzky and Bartley (eds.), *Evolutionary Epistemology*, 47–90, at 59.
[10] W. V. Quine, 'Natural Kinds', in *Ontological Relativity* (Columbia University Press, 1969), 114–38, at 127.

tenth occasion. The somewhat truer, more accurate, faculties here would clearly be disadvantageous in the survival stakes, and one can see a good evolutionary reason for inheriting a somewhat distorted and illusion-ridden picture of the world. Against this example, though, it could be argued that it would be even more advantageous for a creature not to be wrong about predators either negatively or positively, and to be disposed to flee just when there was one near it. Such a creature would waste less effort than the one who fled on nine unnecessary occasions, but would not be eaten either. Truth and usefulness, having been pulled apart at an intermediate stage of evolutionary development, might come together again at a later stage. They might, but, then again, they might not. Not only is there no guarantee that evolution would throw up the refined mutation; morphological considerations might actually render the desired combination of speed and accuracy impossible in the species in question.

To take another example, we can easily show that, on occasion, evolution can produce falsity, rather than truth. A bird may avoid caterpillars with certain types of colouring because they are poisonous; but it will also avoid non-poisonous caterpillars with similar colours, and may be credited with a false belief about the poisonousness of the harmless caterpillar. Of course, the survival chances of the bird are increased by its avoidance of the caterpillar type which includes both noxious and harmless caterpillars. Having a false belief, then, about a particular caterpillar will be a by-product of a survival-producing disposition. Given that the harmless caterpillars have evolved through mimicry of the poisonous ones, we have here an evolutionary explanation of falsehood, reinforcing the general point that there is no direct way of moving from evolutionary workings to truth. As an example of a global error in perception, we might speculate about the world-view of the common frog, who responds visually only to moving objects. A metaphysically reflective frog might then conclude that visibility was necessarily connected to movement: certainly such a belief would be quite consistent with its evolutionarily given and evolutionarily successful make-up, and it is true in its limited perceptual domain, and so in a very rough sense right—but it is not, of course, true generally.

The epistemological lesson to be drawn from these brief remarks on evolutionary theory is that success in the evolutionary struggle considered on its own does not guarantee the truth or adequacy of a creature's beliefs or perceptual representations, even where that creature has been successful in its fight for survival. Its beliefs and representation must be true enough or adequate enough for it to survive, but the 'enough' to promote survival could in theory be consistent with quite high degrees of inadequacy in its beliefs or perceptions. It is in fact always possible to find and refine examples in this area, some apparently showing that truth of belief might be advantageous from the evolutionary point of view, and others apparently showing the opposite. Because of the resulting inconclusiveness, the use of examples here can be at most illustrative.

Some more general thought about belief, theory, and truth will, though, serve to show that its truth or closeness to the truth is only one factor relevant to the usefulness of a belief, and that truth is neither necessary nor sufficient for a belief to be useful from the point of view of survival, or, in more general terms, advantageous. The crucial point here is well made by Campbell and is entirely general:

For rat and scientist, a simple, parsimonious, elegant, few contingencies, few qualifications theory is more useful, more theory-like, of higher decision-rulehood than a more complex one. Without a great deal of parsimony, a theory is not a theory at all, nor a recipe for behaviour a recipe, nor a map of the world a map.[11]

Parsimony, and simplification of data, bias in selection, and distortion in representation are likely to be involved in the representational systems of an organism evolving in a world of any complexity, as ours undoubtedly is. Even if the world were completely ordered—and ours may well not be—what is crucial in the first instance is for members of a biologically produced species to exploit and latch on to any strong approximate orders relevant to their survival, rather than being confused by a welter of apparent disorder or held up by the computations and transformations involved in perceiving the more complex regularities underlying the surface combination of approximate order and disorder. Evolution itself gives us a picture of species evolving in, and being somewhat adapted to, certain fairly specific environments; this adaptation is always seen in terms of the gradual refining of fortuitous and unplanned coincidences or matches between mutations in species and environment. Clearly those matches most relevant to improving survival chances will be latched on to, whether or not they are based in the way the world really is at the level of wavelength spacing, say, or deep causal properties. By the same token, an evolving perceptual mechanism is likely not to latch on to survival-irrelevant regularities which may, nevertheless, be present and causally more fundamental.

[11] D. Campbell, *Descriptive Epistemology* (mimeo.), lecture 4.

In being attuned to the approximate orders of specific environments, the perceptual and cognitive system of an evolutionarily produced species will typically discard information, accentuate aspects of
the information it does abstract, and thereby inevitably give rise to
uncertainties and ambiguities in perception and even, at times, illusions as Floyd Ratliff has shown.[12] Yet, there is clearly going to be a
trade-off between the usefulness of the system in providing cues for
survival-promoting action and its tendency to generate both illusions
and distortions of information more generally. With any information-
processing system at all there are always going to be trade-offs
between speed of processing, goal-relevance of information processes,
and truth of information. When we are thinking of an information-
processing system arising in an evolutionary context, because of the
overall complexity of the world and the environmentally and goal-
specific nature of the tuning filters involved (i.e. the sensory organs
and brains of the creatures in question), it would seem likely on a priori grounds alone that useful and adaptive representational systems
would not in all respects deliver or be seen as aiming at pure truth. It
is likely to be conditioned as much by interplay between specific and
possibly tangential features of the selection domain and the needs and
capacities of the organism. In such circumstances, usefulness of representation could well be compromised by an over-emphasis on the
desirability of unadulterated truth, in so far as the search for truth
could well prevent speed of response, and its acquisition lead to the
suppression of the very simplifications (that is distortions) of data vital
for the guidance of action.[13] On the other hand, as with the frog which
apparently perceives only moving objects in its environment,[14] an illusion-generating perceptual system could produce all the action necessary for the survival of its possessor provided that the illusions in
question were on the whole peripheral to the action-producing perceptual cues, on the edges of the selection domain, so to speak.

Is the situation all that different when we come to reflective human
cognition? As Ratliff correctly notes: '[M]an is probably unique
among animals in that he is aware of at least some of the imperfections
of his sense organs and limitations of his intellect and therefore takes

[12] F. Ratliff, 'Illusions in Man and his Instruments', *Journal of Philosophy*, 68 (1971), 591–7.

[13] Alvin Goldman, *Epistemology and Cognition* (Harvard University Press, 1986), explicitly
introduces speed and goal-responsiveness as epistemic goals, in addition to reliability. He also
says (pp. 124–5) that he is unclear that the various goals sought in cognition can actually be fused
into a single measure.

[14] On this point, cf. my 'Has the Theory of Evolution any Relevance to Philosophy?', *Ratio*,
29 (1987), 16–35, at 28.

steps to overcome them by means of various methods and instruments'.[15] This awareness we have already interpreted as being part and parcel of our self-consciousness, of our conception of ourselves as believers. But, as Ratliff goes on to suggest, the methods and instruments of science, by which we may hope to correct the epistemological imperfections of our sense organs are not themselves illusion-free. Constrast-enhancing devices, for example, generally operate by suppressing low-frequency information. Any frequency modulation device exhibits null points in its output whenever the carrier frequency being modulated corresponds with the frequency of the stimulus input. '[E]very contrivance that is used as an extension of or substitute for (our sense organs) is also a filter, subject to the same natural laws that govern all logical operations, and with peculiarities and limitations of its own.'[16]

Even if we hope to circumvent some of these problems of perception by ascent to scientific theory, we have to remember that in theory there are just the same trade-offs there between system and complexity that we find built into perceptual apparatuses. Nelson Goodman's way of putting the point may seem somewhat extreme when he speaks of the 'sweeping Procrustean simplifications' of science, but it has long been recognized, as he says, that in science (as in any other sphere) 'where truth is too finicky, too uneven, or does not fit comfortably with other principles, we may choose the nearest amenable and illuminating tie'.[17] The moral is that in any cognitive exercise there will be a range of aims, of which truth is only one. Utility of theory or perception may also be very important, so much so that it can and will on occasion partially override or deform truth. This point becomes particularly crucial when we realize that all perception and all theories start from the point of view of living creatures cued into the world at a particular point, and in such a way as to recognize certain features and regularities and to ignore or downplay others which may surround them. Looking at knowledge and perception in an evolutionary context, then, reinforces the distinction between the true and the useful, by showing both that the cognitive inheritance of any creature may be partially determined by a selective and at times illusion-generation attention to its environment which can, at the same time, be advantageous to its chances of survival and reproduction, and also that even in the attempt theoretically to improve its cognitive inheritance, we (or any such creature) are likely to be motivated by criteria for the

[15] Ratliff, 'Illusions', 591.   [16] Ibid. 596–7.
[17] N. Goodman, *Ways of Worldmaking* (Harvester, Hassocks, 1978), 121.

selection of theories which are not the same as, and may even militate against, the search for truth. It does not, of course, follow from any of this that we *should* actually regard our everyday world-view as in some fundamental sense misguided or distorted. I shall, indeed, argue against this suggestion in the next chapter. What, though, I am anxious to establish here is that one cannot move directly from biological success to truth. If one is interested in assessing claims to truth of particular beliefs or systems of belief, the beliefs have to be considered in terms of epistemological criteria such as truth, falsity, empirical adequacy, comprehensiveness, and so on, rather than in terms of utility.

In view of the distinction between epistemological reasons and utility, it is hard to know just how to assess the so-called teleological theory of representation espoused by philosophical naturalists such as David Papineau and Ruth Millikan.[18] According to the teleological theory, the truth-condition for any belief is the condition which guarantees that actions based on that belief will satisfy the desires it is acting in concert with (assuming that any other relevant beliefs are true as well). There is then said to be a biological reason for creatures having true beliefs, so as to ensure that their desires are appropriately satisfied, and it will be in terms of the relevant satisfaction condition that the truth-conditions of belief are given. Papineau admits that in the actual world, false beliefs can satisfy desires, but he says that his analysis allows for this, and allows us to say, on such occasions, that the beliefs are nevertheless false. True beliefs are those which, as he puts it, *guarantee*, for *all* tokens of the relevant types, that ensuing actions will satisfy desires.[19]

But this is of dubious relevance to the operation of biological evolution in the actual world. Evolutionary success is always success relative to specific conditions and actual competitors, and there can never be a presumption that what emerges in evolution will work in conditions or against competitors for which it is not initially selected. In evolution as it is, the functional is always liable to become dysfunctional with very small changes in the situation. It is hard to see how we can base an account of truth on so context-sensitive a quality, given that truth is unchanging. So Papineau is able to fend off the objection about false beliefs satisfying desires only through reliance on a notion of function which is not that of evolutionary theory, but which in its talk of guaranteeing desire satisfaction implicitly refers to what might

[18] Cf. D. Papineau, *Philosophical Naturalism* (Blackwell, Oxford, 1993), ch. 3; R. Millikan, *Language, Thought and Other Biological Categories* (MIT Press, 1984).

[19] Cf. Papineau, *Philosophical Naturalism*, 73.

happen in other possible worlds. And what might or might not be the case in other possible worlds is of no concern to this-worldly evolution, in which this-worldly utility is enough and often more useful than a time-consuming and energy-wasting search for absolute truth or precise accuracy.

When we come to look at knowledge in a social context, it is also very far from clear that truth is necessarily advantageous. Given that societies seek to preserve themselves in existence and that members of those societies have an interest in the preservation of their societies, the deflection of criticism and the avoidance of truth can in certain circumstances be highly adaptive. One reason for this is that one of the central bonding factors in a society is the possession of a shared but particular picture of the world, one peculiar to a specific society by which its members distinguish themselves from other groups and from members of other groups. If a society is partially defined through a system of commonly held beliefs, there is obviously an advantage in having a set of peculiar and idiosyncratic beliefs; up to a point the more peculiar and idiosyncratic, the greater the bonding power, social bonding being highly adaptive for a society in terms of the resulting co-operation between members, yielding in turn such social goods as the division of labour and generational continuity. In the evolution of a society, a false set of community-wide beliefs can be a positive evolutionary advantage. Peter Munz, from whom I derived this point, actually goes so far as to say that 'since "fitting nature" is not the criterion of selection in this case, there is a sort of inverse proportion between being selected and being non-fitting. The more absurd an invention, the more likely its usefulness as a social bond.'[20] Conversely, beliefs which are true are in principle open to anyone to hold, and holding them on rational grounds tends to emphasize one's membership of a universal community of rational agents, rather than one's membership of a specific and demarcated human group. Rationality and conscious, intentional adherence to the canons of rationality can then loosen local and particular social bonds, and it is at least debatable whether or not such loosening is, in a broad sense, socially useful. In more general terms, too, as writers from Plato to Hayek and Roger Scruton have pointed out, there can be great advantages to societies in having mutually advantageous moral practices upheld by myths or noble lies.

Hayek, indeed, has argued that the very customs and traditions which found and underpin the extended orders of market societies

[20] P. Munz, *Our Knowledge of the Growth of Knowledge* (Routledge and Kegan Paul, London, 1985), 292.

were never chosen on rational grounds, nor indeed can they be justified in any direct way.[21] They became established in the first place because the groups which adhered to them became more successful than other groups, and even today we can argue for their preservation only because, and in so far as, we are unwilling to deprive ourselves of the obscure but none the less real contribution such rules make, or may make, to the preservation of the market order. I do not want to say that Hayek's position here is correct, although it is at first sight certainly arguable. (I will have more to say on it in discussing morality in Chapter 6.) It can, though, stand here as an illustration from the realm of morality of our general thesis that the usefulness of a belief or practice may well be distinct from anything perceptible about its rationality. Indeed, the customs Hayek commends on grounds largely of utility may well often be implanted in the community by means of religious myths, a practice he, in common with many conservative thinkers, is loath to condemn.

Nevertheless, there is, as we have seen, a logical connection between belief and truth, and what in the light of Moore's paradox might be termed a logico-psychological drive to arrive at true beliefs and assertions. Even when this drive is overruled by countervailing social and psychological pressures, the distinction between the true and the useful is still part of that concept of belief which itself is part and parcel of our existence as self-conscious agents in a world we conceive as having an existence separate from us. If, though, the notions of truth and usefulness come apart for self-conscious beings when applied to particular beliefs, it might still be argued that the self-consciousness which is the basis of the true–useful distinction is itself useful to survival (rather as rule utilitarians will defend our adherence to specific moral values even while admitting that on occasion doing so may contravene a utilitarian calculation on particular actions). Thus, while it may at times be positively disadvantageous to be saddled with a true belief, the overall advantages of being self-conscious and having a general motivation to aim at the truth greatly outweigh any occasional disadvantages arising from having the concept of truth and of aiming at true beliefs.

Evolution and Consciousness: Advantages and Disadvantages

Attempts to explain the evolutionary advantage of consciousness often focus on what William James called 'variability of behaviour' (in contrast to the tropistic type of response considered earlier). In prac-

[21] Cf. F. A. Hayek, *The Fatal Conceit* (Routledge, London, 1988), ch. 1.

tice, variability entails a system which can (1) assimilate information from the environment, (2) store such information, and (3) use the stored information according to its scheme of preferences. However, while it is clear that our brain and nervous system do exhibit variability in this sense, so it seems do many artificial intelligence machines, and there is no suggestion that they are conscious, let alone self-conscious. To emphasize the role of consciousness in considering responsive awareness of, and flexible response to, our environment actually overlooks the extent to which we often monitor information from our environment and select schemes of action in an unconscious way, for example in driving or games-playing. One's behaviour here is certainly not rigid or tropistic in the manner of creatures whose responses seem unmodifiable in the light of changing situations. In driving and games-playing we do respond flexibly to what is going on around us even to the extent of modifying our strategies and goals, but it is far from clear that we are always conscious of the information being processed or of the goal-modification undertaken. Indeed, to be conscious of all or any of this might well inhibit speedy enough response. Consciousness, in other words, can actually inhibit variability of response, and, in any case, is not necessary for it.

Indeed, rather against the general line of thought which he advocates in *The Self and its Brain*, according to which consciousness is necessary for solving new or non-routine problems,[22] Popper himself follows Medawar in pointing out that the immune system is constantly faced with new problems and frequently solves them, but unconsciously, of course. Popper rather agrees with Dewey (as we shall see) in seeing consciousness itself as sparked into life by an organism being confronted by something unexpected in its environment. He says that only a small fraction of what we learn is admitted to consciousness, and actually recommends psychological experimentation to see which kinds of skill can be acquired only by conscious attention on the part of the learner.[23] In the present state of our knowledge of these matters it is unclear just what contribution to our learning and surviving could only be made by our being conscious. With the possible exception of the addition of 'feel' or 'tone' from our past, as suggested by Edelman, it is hard to think of a *survival-promoting* (I stress) trait or disposition possessed by human beings which might not be possessed by some other non-conscious organism or artefact.

[22] K. R. Popper and J. C. Eccles, *The Self and its Brain* (Springer International, Berlin, 1977), 126 n.
[23] Cf. ibid. 130.

More generally, Popper claims a major evolutionary advantage in letting our hypotheses die in our stead. In other words, just as our sense organs permit vicarious locomotion in the sense that they gather information about places in our environment without our actually having to go to them, so theorizing about the world and testing those theories against the evidence may save us from potentially disastrous encounters with the environment. Popper correctly points out that this formulating and testing of explicit theories requires a representational system, a language of some sort in which the theories are formulated. But the question still remains whether symbolically formulating and testing theories requires self-consciousness. That it may not is suggested by the example of chess-playing computers which do just this sort of thing, in proposing courses of action, and testing and sometimes rejecting what is proposed in the light of information sought or retrieved in advance of trying out the course of action in question. A further question remains about the status of the 'language' or 'representations' involved, because it is hard to see them as semantically engaged with what we take them to be about, precisely because of the lack of consciousness in the machine. But unconsciously and unsemantically, so to speak, the machine that takes in 'information' and vicariously tries out hypotheses before responding to the information would seem to acquire the adaptive advantage we get from consciously proposing and testing theories.

My point here is that many of the supposed adaptive and evolutionary advantages of consciousness and of self-consciousness may be achievable in ways not requiring self-consciousness. This, as just suggested, is certainly true of 'cold' information-processing. It may also be true of many exercises of skill. Indeed, far from unconscious instinctual responses being in tension with conscious and self-conscious learning (as is often claimed), human beings, as William James himself noted, have *both* greater learning capacity *and* more instincts. James's hypotheses have, of course, been given recent support by work on language in the Chomskyan tradition: a language instinct underlying linguistic performance, and being the basis of the learning thus made possible. Moreover, as Matt Ridley has pointed out, Darwin's 'hostile forces of nature' are not a sufficiently challenging adversary for intelligence, or even for consciousness:

the challenges presented by stone tools or tubers are mostly predictable ones. Generation after generation of chipping a tool off a block of stone, or knowing where to look for tubers, calls for the same level of skill each time. With experience each gets easier. It is rather like learning to ride a bicycle. Once you know how to do it, it comes naturally. Indeed, it becomes 'unconscious' as if conscious effort were simply not needed every time. Likewise, *Homo erectus* did not need consciousness to know that you should stalk zebras upwind lest they scent you, or that tubers grow naturally beneath certain trees. It came as naturally to him as riding a bike does to us.[24]

Playing down the supposed advantages of consciousness and self-consciousness for our basic survival against nature does not, of course, mean that as things have evolved a degree of helpful variability of response has not been due to our possession of consciousness and self-consciousness. Nor does it mean, as I shall argue further shortly, that self-consciousness does not give us great potentiality in our competitions with other human beings (who are also self-conscious). What I want to emphasize at this point is that the development of our reasoning powers which has been made possible through self-consciousness has given us cognitive goals which have nothing to do with the acquisition of adaptive beliefs or skills (or with intra-group competition for that matter). The development of our reasoning powers has brought with it the elaboration of a concept of theoretical truth, distinguished from that of the usefulness of theory and belief, and which is impartial regarding usefulness and evolutionary advantage. The reason I say this is because it is never truth-seeking alone which is conducive to survival, but always truth-seeking in conjunction with survival-promoting drives and goals. Our drive to seek truth must at times be subordinated to other goals, even other goals in inquiry, if it is not to impede survival and the speed and direction of response necessary for survival. In other words, a truth-seeking drive and the cognitive structures which promote the realization of such a drive are going to be adaptive and useful to survival only in certain conditions. Only if they do not inhibit speedy enough response, or refrain from overburdening us with irrelevant data, or from making us refuse those very simplifications and distortions of perception and theory which are essential if we are to make our way effectively around the world, will our truth-seeking capacities not prove potentially disadvantageous. In certain conditions, indeed, acting on distorted perceptions or beliefs whose rationale we cannot fully thematize or justify could prove more adaptive than a relentless search for truth. In addition to Hayekian points on the dangers of sweeping away traditional but ill-understood values

---

[24] Matt Ridley, *The Red Queen: Sex and the Evolution of Human Nature* (Penguin, Harmondsworth, 1994), 317.

and institutions, we could cite in this context our innate spacing of sensory qualities and our consequent tendency to project inductively on the basis of these quality spaces. In both types of case, it is clear that in so far as there has been selection of particular belief structures, it has been relative only to particular purposes and particular environments, and that the multiple relativity involved here may make it hard to justify or analyse what has been selected in terms of truth.

One could, indeed, without much difficulty, make the ecological case that our very ability at uncovering truths about nature is likely to promote our downfall imminently, whereas more ignorant and less rationalistically motivated hunter-gatherers survived for millennia in a hard but enduring balance with their environment, largely untroubled, one supposes, by too systematic a search for theoretical truth. On my argument, it would have to be said that these people had not developed as fully as we have the implications of their self-consciousness, but I do not see that this would be an absurd thing to say if one accepts that the development of human powers, including that of self-consciousness, has been a gradual affair, and also one partially dependent on material and economic factors which give one the breathing space, as it were, to develop the powers.

It does seem true, though, that societies in which freedom of inquiry and the search for truth for its own sake are valued have, as a matter of empirical fact, done rather better even in utilitarian terms, than closed societies. After all, it is the Western democracies which have been materially the most successful societies, and it is largely in these societies that scientific research for its own sake has been carried on.[25] We might conclude then, that the search for truth and reason which is an adjunct of our being self-conscious may, for a time, give both individuals and societies an advantage over their more traditionally minded competitors. But there seems nothing inevitable about this. Knowledge of the true causes of things can, of course, be extremely useful if, say, one wants to avoid earthquakes, volcanoes, and hurricanes. But equally, knowledge of atomic and genetic processes, which are also products of our unrestricted desire to know, may not have such beneficial effects, and may, in fact, simply be creating more problems for us, problems which may be beyond our wit to solve. Indeed, some of them, like the problem of nuclear waste, may be insoluble. Equally, our reasoning about goals and ends may lead us to propose to ourselves, as ethically desirable, goals whose attainment may lead to

[25] Cf. J. Marks, 'Ideology and Science, National Socialist Germany and the Soviet Union', *Salisbury Review*, 4 (1986), 2–7.

the decline of individuals and species. This, I take it, is part of the burden of complaint (which we will examine further in Chapter 6) made by some evolutionists against a morally desirable ethics of co-operation, on the grounds that struggle and competition are the essential engines of the material welfare which is the basis of a reasonable existence for all; again, a problem that is posed by our reasoning, in this case about ends, but which may be rationally insoluble.

More generally, there might be something in Nietzsche's complaint that reflective reasoning in general, and the scientific pursuit of truth in particular, might be life-denying *per se*, precisely because they have a tendency to produce in those subject to their exigencies a tendency to promote an 'enlightened' but life-denying attitude to life, and to downgrade more robust and life-enhancing dispositions.[26]

It seems to me, however, that global arguments about the usefulness of our powers of reasoning about truth and goals are likely to be and to remain inconclusive, and this is really all I want to maintain at this point. They are likely to remain inconclusive precisely because there is no necessary connection between what is true and reasonable and thought to be desirable, on the one hand, and beliefs which might be practically useful to have, on the other. By virtue of our existence as self-conscious agents, we have some imperative both to admit the validity of the distinction and, in matters of belief, to seek the true rather than the merely useful, and in matters of desire to seek the good rather than the merely useful.

The point I have been making about the distinction between the true and the useful in matters of belief can be brought out in a complementary way by contrasting the role of belief in self-conscious (presumably human) agents and of belief in animals. As we have seen, we can think of animals, such as horses or dogs, having beliefs in a certain sense: that is, their behaviour is guided by consciousness or awareness, for example, that a certain fence is of such a height at such a distance. They are conscious of their environment, and they act on the consciousness of it they have. Obviously it will be advantageous to animals to have beliefs which help them cope successfully with their environment. Often such beliefs will be true beliefs: for example, a horse believing that a fence is four feet high as opposed to six feet (however that is represented to a horse) will not help the horse if the fence really is six feet. So, it might be argued, against the tenor of my

[26] Cf. F. Nietzsche, *The Genealogy of Morals*, 1887 (Modern Library, New York, 1986), pt. 3, sect. 25.

argument linking the truth-drive with self-consciousness, that there is already in unselfconscious animal belief a teleology towards truth.

Against this, though, I will insist that if there is such a teleology it is a somewhat accidental one. True beliefs may, on occasion, be pragmatically useful to a horse or a dog. But there is no general requirement that useful beliefs are true or that true beliefs are useful. The drive or teleology in the conscious but not self-conscious animal is towards perceptions and beliefs (mental representations if you like) which are useful, which help it cope with its environment, and such beliefs and perceptions can be distorted or even false. As we have seen, a creature lacking language and self-consciousness lacks the ability to conceive of its perceptions and beliefs as possibly false, precisely because it lacks the conception of itself as a believer at all, even though in a perfectly natural sense it may be said to have beliefs. And lacking the conception of itself as a believer, it will lack just that teleology towards the truth which characterizes self-conscious believers, the only ones who can make the distinction I am interested in, between true beliefs and useful beliefs.

In the light of what I have been saying about the connection between self-consciousness and the disinterested drive to the true, and about the possibly problematic connection between the possession of self-consciousness and evolutionary advantage, it might be worth mentioning the view that our drives to disinterested cognitive and aesthetic activities—and perhaps our self-consciousness itself—are by-products of our extraordinarily lengthy infancy and youth: that unlike most other species, human young are born immature with others perforce satisfying their instinctual needs. In the metaphorical space thus given, human beings can then develop aims and capacities of a disinterested sort which have in themselves no survival potential. But why are human young born so immature? There is evidence that if early men had had brains larger and more developed than ours at birth, up to 90 per cent of childbirths would have ended fatally for mother and offspring. Our being born more immature and less developed thus helped childbirth, and hence contributed to the reproductive success of *Homo sapiens*. But if our development of a disinterested sense of truth and beauty are by-products of our being born so immature, our immaturity also had consequences of immense importance for the future development of our species.

Dewey and Naturalism

I now want to look briefly at the account of knowledge given by John
Dewey, because he argues explicitly for what should certainly be
thought of as a biological analysis of knowledge. That is to say, he sees
the acquisition of knowledge in specifically biological and evolutionary terms. In general, he sees thought and speech as phenonema produced by the environment by means of organic structures. More
specifically, conscious thought arises only when there is some im-
balance between a human organism and its environment, when some
settled habit or disposition is unsettled. 'The disturbed adjustment', as
he puts it, 'of organism and environment is reflected in a temporary
strife';[27] this temporary strife is what actually produces conscious
experience and thought, by contrast with the subconscious activity of
mind which is both enormously wider than that of consciousness and
persistent. 'Consciousness is, as it were, the occasional interception of
messages continually transmitted',[28] while mind is a constant background, the source of a constant and habitual interaction between
organism and environment.

Once consciousness has been sparked by some disturbing impulse,
such as 'the blocking of efficient overt action',[29] equilibrium has to be
re-established. The now fragmented harmony between the various
elements of the situation, as it appears to the organism, have to be con-
verted into a unified whole in which once again there will be an
absence of felt conflict and a reappearance of settled habits—until, of
course, these new habits are challenged by some further discordant
experience. Cognitive validity is defined in terms of the consequences
of acting on an idea, where these consequences fit harmoniously into
the consequences of acting on other ideas. Scientific inquiry is the
activity of perceiving the eventual agreement or disagreement of such
sets of consequences.[30] For Dewey, the most worthwhile and meaningful experiences are those with the greatest links with other experiences, the greatest unificatory potential, the greatest potential for
further and more complex types of experience.

Dewey's account of knowledge is not only biological in structure. It
also glosses over the very distinction between the true and the useful
(or experientially satisfactory) which we were labouring over in the

---

[27] J. Dewey, *Human Nature and Conduct* (Modern Library, New York, 1930), 179.
[28] J. Dewey, *Experience and Nature* (Dover Publications, New York, 1958), 303.
[29] Dewey, *Human Nature and Conduct*, 190.
[30] Cf. Dewey, *Experience and Nature*, 323–4.

last section. And this, I believe, is a direct consequence of the adoption by Dewey of a biological analysis of knowledge. Just because Dewey takes knowledge to arise only in the context of the interruption of habits of action and to consist in the resolution of such interruptions, he is unable to account for the truth-seeking aspects of scientific activity, in which many problems do not arise (except in a vacuously extended sense) from interruptions to activity, and are not solved by an organism acquiring a less defective set of habits. Yet such truth-seeking is part and parcel of our conception of ourselves as existing in an objective world.

Dewey's conception of the value of experiential growth is also hard to reconcile with his analysis of inquiry (and so, incidentally, is his advocacy of continuous reconstruction of habits). For if what stimulates inquiry is what unsettles us, should we not rest content with a settled state, where exposure to new experiences might be unsettling? Nevertheless, one can see why Dewey might have been led by his evolutionary model to his advocacy of experiential growth, and this for two reasons. The first of these goes back to the first and third characteristics of biological evolution mentioned at the start: the idea that survival depends on the ingestion of food from the environment, together with the corollary of living things creating scarcity by virtue of the mere fact of living. For a survivor, survival and growth, or at least development, go hand in hand; a living thing requires ever new inputs of food and other life-sustaining elements such as air. If we are looking at knowledge in these terms, it might well seem that continuing inputs of new knowledge are essential for the survival of a knowing organism.

Then, second, there is the question of evolutionary progress. The very word 'evolution' suggests that in life there is development in the direction of more complexity in species, more finely balanced attunements of organisms to environment, more subtlety and sophistication in organs and dispositions. It must be remembered here that Darwin himself was cautious about the use of the term, generally preferring the drier phrase 'descent with modification' to the progressivistically nuanced 'evolution'. Strictly speaking, within the terms of the theory of evolution, the more evaluatively neutral standpoint is correct. Darwinian evolution and the law of natural selection do not entail progress in any absolute sense, but only a higher degree of fitness relative to a given environment than is possessed by competitors in the same environment. This does not imply an increase in faculties or organs or in complexity within a successful species; if providing simpler organization and shedding of faculties or organs produces a greater adaptation to the environment in which a species finds itself, these things can and do occur within natural selection. Nor, as Darwin was well aware, does natural selection imply a high degree of fit between species and environment, but it is tolerant of wastefulness and lack of fit provided no better fitting competitors appear to exploit the mismatch which lies between organism and environment.

[A]s natural selection acts by competition, it adapts the inhabitants of each country only in relation to the degree of perfection of their associates; so that we need feel no surprise at the inhabitants of any one country, although on the ordinary view supposed to have been specially created and adapted for that country, being beaten and supplanted by naturalised productions from another land. Nor ought we to marvel if all contrivances in nature be not, so far as we can judge, absolutely perfect; and if some be abhorrent to our ideas of fitness. We need not marvel at the sting of the bee causing the bee's own death; at drones being produced in such vast numbers for one single act, and being then slaughtered by their sterile sisters; at the astonishing waste of pollen by our fir-trees; at the instinctive hatred of the queen bee for her own fertile daughters; at ichneumonidae feeding within the bodies of live caterpillars; and at other such cases. The wonder indeed is, on the theory of natural selection, that more cases of the want of absolute perfection have not been observed.[31]

Nevertheless, against the austere 'descent with modification' version of Darwinism, it should be pointed out that Herbert Spencer, whom Darwin much admired as a philosopher and expounder of the theory of evolution, definitely propagated the idea that evolution was a progression from 'low' to 'high' types, and that such a progression was necessary both for the continuance of life and for increase of happiness. Darwin, too, was happy enough to speak of each step in natural selection as implying an 'improvement' in a species 'in relation to its conditions of life'. In the very last paragraph of the *Origin of Species* Darwin speaks of 'the production of the higher animals' as directly following from 'the war of nature, from famine and death', concluding

there is a grandeur in this view of life with its several powers, having been originally breathed into a few forms or into one; and that, whilst this planet has gone cycling on according to the fixed law of gravity, from so simple a beginning endless forms most beautiful and most wonderful have been and are being evolved.

[31] C. Darwin, *The Origin of Species* (Penguin, Harmondsworth, 1984), 445.

This wording does suggest a more Spencerian perspective, even to the extent of using the very term 'evolved', despite the strictly logical requirements of the theory of evolution. And, as we will see further in Chapter 6, in *The Descent of Man*, Darwin is quite confident that our development as a species is in the direction of greater virtue and genuine altruism towards our fellow men (though how consistent this is with his views on the war of nature being the wellspring of development we will have to consider). Darwin also wrote in a letter that in 'looking at the world at no very distant date . . . an endless number of the lower races will have been eliminated by the higher civilised races throughout the world'.[32]

Dewey held a view of nature which—as in all his thinking—emphasized continuity rather than discontinuity. While he distinguished between the physical, the psycho-physical, and the mental, these distinctions are not to be construed in terms of differences of stuff or even of types of impulse, so much as in terms of differences of complexity and 'intimacy of interaction among natural events'. Indeed, while he thinks that a merely physical thing such as an iron molecule is, in terms of integration and equilibrium, different from an animate plant, and the plant different again from the human being, 'it is reasonable to believe that the most adequate definition of the basic traits of natural existence can be had only when its properties are most fully displayed'.[33] In other words, we understand the physical fully only by reference back from the mental. In mental activity the active relationship of organism and environment is understood and seen for what it is by being taken up through language in a shared system of signs. What Dewey calls growth and complexity are most present when a material being not only participates in the natural work as iron molecules no doubt do, but does so with both sentience (psycho-physical level) and understanding of the relationship between itself and parts of the world and other similar experiences (mental-linguistic level). This account of the relationships between different levels of existence can clearly be seen in evolutionary terms. It also, for Dewey, evolves into morality:

[T]he process of growth, of improvement and progress rather than the static outcome and result becomes the significant thing. . . . The end is no longer a

[32] C. Darwin, *Letters*, vol. i, ed. F. Darwin (London, 1887), 316. (I derive the picture of the progressivist in Darwin from Roger Trigg's chapter on Darwin in his *Ideas of Human Nature* (Blackwell, Oxford, 1988). Trigg correctly notes there that neither genuine altruism nor talk of races in the intra-human context are easy to reconcile with the strictly biological account of genetics. Both are realities that arise only in human society.)

[33] Dewey, *Experience and Nature*, 261–2.

terminus or limit to be reached. It is the active process of transforming the exist-ent situation. Not perfection as a final goal, but the ever-enduring process of perfecting, maturing, refining is the aim of living. Honesty, industry, temper-ance, justice, like health, wealth and learning, are not goods to be possessed as they would be if they expressed fixed ends to be attained. They are directions of change in the quality of experience. Growth itself is the only moral 'end'.[34]

In other words, what is moral is what promotes evolutionary growth, in Dewey's sense; that is, change in the quality of experience wrought through mental processes such as inquiry, co-operative social activity, and aesthetic experience. All these things move us away from dis-ordered, meaningless experience in the direction of harmony and meaning.

If one accepts a progressivist reading of evolution, Dewey's account of inquiry and mental activity more generally is extremely close to what one might expect from seeing them as natural processes: they are seen as processes whose purpose is to establish harmony between organism and environment at the level of the mental, without neces-sarily implying any precise fitting of organism and environment or mirroring of one by the other. But even if the progressivist reading of evolution is rejected, Dewey's account of the mental can still be seen as naturalistic, in its stress on such notions as order, unity, and har-mony, and his systematic refusal, under questioning from Russell and others, to make any distinction between the true and what inquirers might see as meaningful, as a solution to problems or disturbances confronting them.[35]

In this, of course, Dewey repudiates the very distinction we have been labouring to maintain; while this is an unsustainable refusal, given our self-conscious, reflective nature, the fact that so dedicated a naturalist makes it is, in its way, a vindication of our view that natu-ralism renders the crucial distinction hard to account for.

Peirce and the End of Inquiry

Despite being generally seen as the founder of pragmatism—or, prag-maticism, as he later came to insist, partly because of misunderstand-ings thereby engendered—a rather different perspective on knowledge

---

[34] J. Dewey, *Reconstruction in Philosophy* (Boston, 1948), 177.
[35] See particularly Dewey's 'Propositions, Warranted Assertibility and Truth', in his *Problems of Men* (Philosophical Library, New York, 1946), 343–4. As such, it would not be altogether mis-leading to speak of Dewey's account of knowledge as Darwinian.

and evolution from that of Dewey is presented by C. S. Peirce. Along with Dewey, Peirce did, it is true, adopt a broadly verificationist account of meaning, in terms of the experiential difference a word or concept might make. Like Dewey he also sees inquiry as a process arising from the irritation of doubt, and belief being a state which soothes the irritation. But these similarities cannot hide great differences between the two thinkers, and nor should their common espousal of what might broadly be called an evolutionary perspective.

Dewey is all for experience, and as much sharing of it and as many interrelationships as possible. He spoke disparagingly of 'what is termed spiritual culture' as something 'rotten' because 'it has been conceived as a thing which a man might have internally—and therefore exclusively'.[36] He also wrote of the 'intrinsic significance of every growing experience',[37] and assessed the worth of a form of social life in terms of the extent to which the interests of a group are shared by *all* its members, and in its ability to communicate to other groups. In contrast to this democratic sentimentalism, Peirce appears distinctly crabby and curmudgeonly. No reader of the journal *Science*, he says,

> is likely to be content with the statement that the searching out of the ideas that govern the universe has no other value than that it helps human animals to swarm and feed. He will rather insist that the only thing that makes the human race worth perpetuation is that thereby rational ideas may be developed, and the rationalisation of things furthered.[38]

For Peirce, in direct contrast to Dewey, knowing rather than being or doing is the chief end of the university, if not of life itself. It is easy to see why in an increasingly utilitarian climate he came to despair that he had spawned a philosophy called pragmatism, a philosophy of which Dewey was to become a notable and quite un-Peircian exponent.

Peirce, like Dewey, is an evolutionary thinker. He constantly stresses a movement in the universe to increased reasonableness. It is clear that by this he does not just mean increase of complexity and order. It does mean this but, over and above increased complexity and order, it is awareness of this order which Peirce is interested in. Peirce would not, I think, dissent from the idea, encapsulated in some strong versions of the so-called anthropic principle, that the physical universe is an integrated physical system which (in the words of Jay Rosenberg)

---

[36] J. Dewey, *Democracy and Education* (Macmillan, New York, 1961), 143.
[37] Ibid. 109.
[38] C. S. Peirce, 'Review of Clark University, 1889–99', in *Science* (1900), 620–2, quoted in *Charles S. Peirce: Selected Writings*, ed. P. Wiener (Dover Publications, New York, 1958), 333–4.

'necessarily "grows knowers" and which thereby comes to mirror itself'.[39]

The strain of idealistic thinking in Peirce's thought is actually rather stronger than mention of the universe as a physical system and of the anthropic principle might suggest. Proponents of the anthropic principle tend towards some version of materialist monism, or at most to a dualism of material substance and emergent psycho-physical properties. Mental activity, on this view, is created by the material universe as a reflection of the universe, which explains the tendency of such activity to mirror the universe. We can come to understand the universe in part by considering our dispositions to such mental activity and the circumstances necessary to bring such activity about. In the strong version of the principle, having discovered what these circumstances are we come to see that there is something in the nature of fundamental natural processes which is bound, under certain circumstances, to bring about conscious mental activity.

By contrast, Peirce's idealism involves a monism of mind. He rejects the idea of two fundamental substances, mind and matter, finds Newtonian accounts of the genesis of sentience unsatisfactory, and concludes that all that really exists is mind.[40] Against this background, we can make sense of Peirce's faith in our powers of abduction (plumping naturally for the likely hypothesis about how things are), in our natural instinct for truth, and in the ultimate convergence of disinterested inquirers on *the* truth. Being part of the universal mind, we have, as he says, 'some glimmer of co-understanding with God, or with Nature'.[41] Indeed, we are part of God, part of the great universal mind, which is gradually growing in complexity and self-understanding although, to be sure, he is aware of the difficulties involved in speaking of the perfect God growing and developing, difficulties which process theology continues unavailingly to grapple with.[42]

So, as Hookway points out,[43] Peircian 'agapeistic' evolution, driven by divine purpose and love, is not Darwinian natural selection with its blind and wasteful mechanisms, and its lack of guarantees regarding fit and precise mirroring. Peirce, like Darwin's own collaborator Alfred

[39] J. Rosenberg, *Linguistic Representations* (Reidel, Dordrecht, 1974), 144.

[40] Cf. C. Hookway, *Peirce* (Routledge and Kegan Paul, London, 1985), 275.

[41] Cf. C. S. Peirce, *Collected Papers*, vol. viii (Belnap Press, Cambridge, Mass., 1935), para. 211.

[42] Cf. C. S. Peirce, *Collected Papers*, vol. vi (Belnap Press, Cambridge, Mass., 1935), para. 466. On process theology and its problems, cf. my *Experience, Explanation and Faith* (Routledge and Kegan Paul, London, 1984), 185–200.

[43] Hookway, *Peirce*, 280.

Russel Wallace, was well aware that the Darwinian war of nature and struggle against famine and death could do little to explain our higher intellectual faculties, our impersonal morality, or our aesthetic sense in its full disinterestedness. Both Peirce and Wallace moved in the direction of religious solutions to account for our higher powers and capacities. Even today, our disinterested pursuit of truth and our actual success in delving into such non-utilitarian matters as higher mathematics and astrophysics are commonly perceived as a problem by philosophers who do not on the face of it have religious or theological interests.

The argument about theoretical knowledge is put most starkly by Thomas Nagel: 'if, *per impossibile*, we came to believe that our capacity for objective theory were the product of natural selection, that would warrant serious scepticism about its results beyond a very limited and familiar range'.[44] The point is that if our mental and perceptual faculties have evolved in natural selection, they then would be adapted only for circumstances similar to those for which they had evolved; dealing with predators on the savannah, hunting and gathering in similar conditions, and so on. We clearly do not need poetry or music or quantum theory or astrophysics or Gödel's theorem or even speculative philosophy for any of that. So why should we expect instruments honed for the one purpose to be any good for the other? Nagel concludes that the development of the human intellect might be taken as 'a probably counter-example to the law that natural selection explains everything'.[45]

Nicholas Rescher is not as severe as Nagel, arguing that our intellectual ability 'to pursue various challenging projects that have nothing to do with survival'[46] is needed to help solve ordinary survival problems very quickly. Nevertheless, he speculates that our science can never do more than give us pictures of the world constrained by what he calls a 'particular evolution-determined mode of emplacement in the world's scheme of things'.[47] Our science might be incommunicable to beings with a different orientation to the world, and vice versa.

Michael Ruse, a thoroughgoing Darwinian, accepts what he takes to be the implications of Darwinism regarding our mentality, and is quite prepared to live with the consequences. Our science, our logic, and

[44] Nagel, *The View from Nowhere*, 79.
[45] Ibid. 81.
[46] N. Rescher, *A Useful Inheritance: Evolutionary Aspects of the Theory of Knowledge* (Rowman and Littlefield, Savage, Md., 1990), 110.
[47] Ibid. 77.

our mathematics are all based on principles innate in us, but adapted because of their survival value. Our scientific theories could be no more than 'illusion fostered [*sic*] on us for reproductive purposes'.[48] Similarly the necessity we find in logic and mathematics may be a device for reinforcing logical and mathematical beliefs: 'Biological fitness is a function of reproductive advantage rather than of philosophical insight. Thus if we benefit biologically by being deluded about the true nature of formal thought, then so be it. A tendency to objectify is the price of reproductive success.'[49] (As we will see, Ruse has similar things to say about 'illusions of objectivity' in morality.)

There is certainly something intriguing in what might be called the sceptical naturalism of Nagel, Rescher, and Ruse, and aspects of it are hard to dismiss, as we shall shortly see. However, as Michael Smithurst (on whose work I have drawn in the last few pages) points out, Darwinians are not left as bereft of resource in this area as, in their different ways, Nagel and Ruse at least imply.[50] For Darwinism should not be seen as a theory simply about survival against nature, famine, and natural death. It is also about reproductive success against one's con-specifics. For peacocks, as we shall see, competition for the favours of peahens means ever more elaborate tails. For human beings, competition becomes increasingly a tournament of the mind, a matter of second-guessing and outwitting our fellows. Matt Ridley, as we have seen, argues that the more skilled we become at subsistence survival, the less conscious we need be: nature on her own does not throw up new challenges that often. By contrast, human beings throw up ever new challenges to each other: 'the faster mankind ran—the more intelligent he became—the more he stayed in the same place, because the people over whom he sought psychological dominion were his own relatives, the descendants of the more intelligent people from previous generations'.[51]

So, we can give an evolutionary explanation for the emergence of creative intelligence and imagination. But, as Smithurst implies, creativity is quite a different thing from a mundane and narrowly focused skill. A capacity to crack walnuts could have been selected for, but it just makes no sense to speak of a capacity for astrophysics being selected. What may have been selected was the underlying intelligence

[48] M. Ruse, *Taking Darwin Seriously: A Naturalistic Approach to Philosophy* (Blackwell, Oxford, 1986), 188.

[49] Ibid. 172.

[50] Cf. M. Smithurst, 'Popper and the Scepticisms of Evolutionary Epistemology, or, What Were Human Beings Made For?', in *Karl Popper: Philosophy and Problems*, 207–24.

[51] Ridley, *The Red Queen*, 321.

and flexibility of mind, and that selected for the competitive edge it gave its possessor.

After arguing that instinct would do better at most of the tasks normally assigned to consciousness and intelligence, Geoffrey Miller sums up the position as follows:

The neocortex is not primarily or exclusively a device for tool-making, bipedal walking, fire-using, warfare, hunting, gathering, or avoiding savanna predators. None of these postulated functions alone can explain its explosive development in our lineage and not in other closely related species. . . . The neocortex is largely a courtship device to attract and retain sexual mates: its specific evolutionary function is to stimulate and entertain other people, and to assess the stimulation attempts of others.[52]

Smithurst is more blunt, alluding in addition to the role of deception in our competitions with each other: human beings are made, he says, 'to tell lies and have sex'.[53] More precisely our specifically human self-consciousness and intelligence is a function of our competitions with each other for sex.

What all this shows is that we do not need to look at our mental faculties in terms of mirroring specific environments. While our brains did not evolve to do astrophysics, they did not evolve just to get us around the Rift Valley either. Some of the negative and restrictive conclusions of Nagel and Ruse are not called for. At the same time, however, it cannot be said that everything they and Peirce and Wallace are concerned about has been dispatched as a non-question.

Even if it were true that higher mathematicians, astrophysicists, and quantum theorists were mainly motivated, consciously or unconsciously, by the sexual rewards they might be brought by their efforts, we would not have gone one step in showing just why it was that efforts of these sorts were admired by those in a position to provide the rewards. We would not have shown why human ingenuity was directed into these efforts, nor more generally why the pursuit of truth had come to be seen as a good.

It has been the theme of this chapter to suggest that self-consciousness brings with it the reflectiveness to raise questions about truth, though even saying this does not explain why the search for truth has come to play so large a role in human affairs. Peirce's ques-

[52] G. F. Miller, 'Sexual Selection for Protean Expressiveness: A New Model of Hominid Encephalization', paper delivered to the 4th annual meeting of the Human Behaviour and Evolutionary Society, Albuquerque, New Mexico, 22–6 July 1992, quoted in Ridley, *The Red Queen*, 327.

[53] Smithurst, 'Popper and the Scepticisms of Evolutionary Epistemology', 212.

tion about the rationalization of things has not simply evaporated in our recent thoughts about the means by which some humans come to swarm and breed more than others.

Nor, indeed, though this whole chapter has been on the topic of evolutionary epistemology, have we actually gone very far in reaching any epistemological conclusions. We have said a lot about the origins of faculties and beliefs, but little about their status, or justification, beyond suggesting that naturalistic accounts cannot be used to side-step such questions. In this chapter, my negative conclusions have derived largely from an analysis of attempts to use the theory of evolution as a means of epistemological justification. In the next chapter, I will attempt to derive some more positive results, paradoxically perhaps, from evolutionary pessimism regarding our fundamental world-view.

5

Evolution and Epistemological Pessimism

Epistemological Pessimism

Even if, as we argued in the previous chapter, nothing directly justificatory about our beliefs can be derived from our success in biological evolution, it does not follow that in looking at man as a product of evolution, nothing of philosophical interest can be said about our systems of belief and perception. Indeed, rather against the optimistic tone of much evolutionary epistemology, including Lorenz's, and elaborating some of the themes we adduced in criticism of Lorenz, one might be tempted to draw a pessimistic or sceptical inference along the following lines. Each species which survives for any length of time has to consist of individuals able to survive to maturity and reproduce. Where it is relevant, this means that their cognitive faculties have to be reliable enough to allow them to survive and reproduce. But being reliable enough in this sense is quite consistent with considerable degrees of inaccuracy and unreliability and, above all, with considerable selectivity and partiality of perspective. (Think of the frog's vision; think of the deafness of snakes to airborne sound; think of our own unresponsiveness to sonar.) In so far as evolution has anything to teach us epistemologically, it is that we should treat our initial perceptions and beliefs with caution. We should regard them as likely to be partial and selective, and in some cases even false. As we saw with Rescher and Ruse, evolutionary theory emphasizes the multiplicity of viewpoint, the prodigality of cognitive ecological niches, and above all, the need to break out of one's initial world-view if one is interested in a wider and more inclusive picture of the world.

It is indeed in this way that Bernard Williams has understood the role of evolutionary biology in a theory of knowledge.[1] Biology, together with the neurological sciences, will provide non-vacuous explanations of how creatures, including human beings, come to have

---

[1] Cf. B. Williams, *Ethics and the Limits of Philosophy* (Fontana, London, 1989), 139–40.

the perspectives they hold on the world. The clear implication is that biology and neurology will show us that, in certain respects, the cognitive perspectives we and others have are partial and misleading. No doubt they will show that in other respects these perspectives are correct, but the role of the life sciences is here being conceived as critical rather than justificatory. They are concerned with elaborating and explaining the contrast between a partial perspective on the world and one which, ideally, will be absolute, freed from any particular perspectives and modes of perception. The theory of evolution then, can be seen as encouraging an epistemology which will lead us to take up a critical attitude to the modes of perception and ways of thinking we are born with.

In doing this, an evolutionary approach is likely to lend support to the doctrine characteristic of many scientific or naturalistically inspired epistemologies: the distinction, that is, between primary and secondary qualities. It will do this because it will show first in some detail just how the perception of secondary qualities is produced in the organism. Then, in distinction to primary qualities it will be said, secondary qualities do not represent things as they really are. Locke, in his classic discussion of primary and secondary qualities, says that whereas primary qualities are resemblances of bodies, secondary qualities do not resemble the things that cause them; they are nevertheless caused by things, apart from us, and vary by and large according to variations in their causes. Biology and neurology show us how they are produced by interaction between things and us. Thus, a study of the biology of vision will show us, in the words of C. L. Hardin, that 'coloured objects are illusions, but not unfounded illusions'.[2] This contrasts with our perceptions of the shapes of things which (according to proponents of the primary–secondary quality distinction from Locke to Hardin) do have their analogies in the real world.

All this sits well with what, following Campbell and Ratliff, as well as Rescher and Ruse, we may have concluded from the theory of evolution. The perception of secondary qualities is a useful exploitation of certain interactions between wavelengths emitted by objects and our perceptual faculties. It is useful because it produces in us experiences (of colour, sound, and the rest) which have survival value for us. Perceiving things as coloured, noisy, and so on is useful to us in coping with reality. But outside the perceptual niche occupied by humans and similarly endowed creatures, things do not appear to have such

---

[2] C. L. Hardin, *Colour for Philosophers* (Hackett, Indianapolis, 1988), 111.

qualities. Even inside it, their appearing to have them is due to our sense organs as much as to the things themselves. (As Hardin has shown, there is no set of microstructural physical characteristics of the objects themselves which determine their colours, nor is any given colour exclusively associated with specific spectral emissions.) Secondary quality perceptions are an evolutionary artefact, in no way true to the way things really are. If it seems surprising that we could be misled by our biological make-up into thinking the world has all sorts of properties it does not really have, this is only because we have forgotten that evolution guarantees us at most survival-permitting faculties. To think that they are thereby truth-delivering is misguidedly to assimilate the survival-permitting in cognition to the true.

An evolutionary perspective on secondary qualities, then, may seem to lead to the conclusion that they are unveridical, and that, therefore, a large part of our world-picture is projection on our part. We project colours and the like on to things which are really colourless, silent, etc. Evolution, then, leads to scepticism or at least to doubt, rather than to epistemological assurance. (Remember that the theory of evolution is as much about lack of fit, as much about bodging and the exploitation of random coincidences of organism and world as about super-intelligent design.) What I now want to suggest is that we need not draw such negative conclusions from evolution, but before doing this I need to clarify just what follows from the anti-justificationist reasoning of the last chapter. That argument, it will be remembered, was directed against Lorenz's claim that modern science in general, and biology in particular, could provide some answer to scepticism and Kantian idealism.

In order to refute this claim, we saw that the need for biological survival *could* motivate perceptions and beliefs which, from a scientific point of view, were distorted and partial. But it does not, of course, follow that our fundamental everyday picture of the world is, in some serious sense, wrong. That would be so only if we were claiming that the scientific point of view itself were more correct than the everyday. To assume that it might be was legitimate in discussion of Lorenz and standard evolutionary epistemology, for the premiss of such arguing is that it is (as it is equally the premiss of the naturalistic pessimism we are now examining). What in the case of Lorenz I was anxious to counter was the thought that there could be any quasi-scientific refutation of scepticism, based on the theory of evolution. What is actually the case is that, in a certain sense, the theory of evolution, considered as an account of our survival as a species, is epistemologically neutral.

Considered as an account of our survival and success through time, it licenses neither optimism nor pessimism about our everyday view of the world, and this is partly because survival and reproductive success do not bear directly on the truth or otherwise of any world-view or set of beliefs. However, what I will now suggest is that we can think of biology and evolution as conveying rather more than the purely Darwinian message that we have survived, and that looking at it in this broader way, and ourselves as creatures living in and reacting to a real world can, despite Hardin and Williams, actually ground our primitive sense that the world is filled with secondary qualities. I will proceed by first suggesting some general reasons for rejecting scepticism about secondary qualities, before moving on to sketch the account of knowledge which evolution in its broader sense might suggest.

Science and Secondary Qualities

A first reply to scepticism about secondary qualities is to point out that sauce for the secondary goose is sauce for the primary gander too. In other words, the perception of shape, extension, movement, and other primary qualities turns out to be as dependent on the states of perceivers as the perception of secondary qualities. This criticism of Locke dates back to Berkeley, who pointed out that secondary quality perception was intimately involved in the perception of primary qualities. (That is, you cannot observe the triangularity of an object without observing the colour of its surfaces.) More recent psychological data suggest further that the perception of shapes and movements depends essentially upon nervous systems, and is as open to distortion and ambiguity as the perception of colour and the other secondary qualities. We learn, moreover, from relativity theory that our perceptions of primary qualities such as velocity and extension are also determined by our positions and velocities relative to those of the objects we are observing. In other words, when it comes to the manifest image—what we actually perceive—primary qualities are as observer-dependent as secondary qualities. From the point of view of any possible experience, the absolute and invariant 'intervals' between spatio-temporal events postulated by the theory of relativity are merely an abstraction from all points of view, and were regarded as such by Einstein. They are not something more certain and more objective than any of the perspectives from which they are abstracted. The difference between primary and secondary qualities would then seem at best a matter of degree. Both are caused by observer-independent objects; neither represents objects independent of any point of view, but what we perceive of the so-called primary qualities may be closer to things in themselves than what we perceive of their secondary qualities.

Actually, I doubt very much that this is the case, if by things in themselves we mean what is delivered by the scientific image of contemporary physics. This image is not of the everyday world rendered quiet and colourless, as in silent, black-and-white film. From the account of my room given by modern physics we should find it hard to recognize the objects of daily experience even by means of their primary qualities. The solid, clearly defined table on which I write is in the physicist's account altogether more fuzzy and gappy, and far less easily demarcated from its surroundings. Doubtless there are lattices of particles which underlie the table and cause my perceptions of it, but their extensions, motions, and configurations in the scientific image are very different from my perception of the table's extension, movement, and configuration. Unless we read the table back onto the region of physical space-time which it occupies, it is not at all clear that its primary qualities would leap out at us from the scientific image as constituting a definite and discrete object in that image, clearly unified in itself and separate from the things on and around it. The point I am making here is similar to the familiar Wittgensteinian point about the broom: that in the absence of our use of brooms, we would have no clear sense of the object in the cupboard being one thing rather than two or many. I suspect that when the scientific image is taken as primary, the identity of the everyday objects of the manifest image—and hence their primary properties—will be far from evident. The resemblance between them and their primary properties and what is given in the scientific image will be accordingly distant. More generally the point is that the identity of objects, and hence the consequent synthesizing of particular primary properties into objects, is not given to us independent of particular practical interests.

The scientific account of the world, though, suggests one respect in which primary qualities are more fundamental than secondary qualities. For there is no variation in anyone's secondary quality perception without a variation at the primary quality level. There is, in addition, an assumption that there are or will be found lawlike causal explanations at the primary, but not at the secondary level. So, even if, as Hardin says, no independently specifiable set of microstructural properties corresponds to any given secondary quality, secondary quality perception supervenes on the causally more basic and nomologically more satisfactory primary qualities.

Apart from psychological states then, we can causally explain everything which happens in the world without invoking the secondary qualities which perceivers perceive. Further, these perceptions themselves depend on interactions between primary qualities of things and our sense organs, even if not in a perspicuously lawlike way.

Thus, from the point of view of a certain explanatory project, some properties are more fundamental than others. This might seem once more to imply as Hardin has it that the more fundamental properties are objective (or more objective), while the secondary properties are subjective illusions. There is nothing in what the physicist tells us about the world which precisely corresponds to our division of the world into colours: further our colour perception supervenes on the causally more basic qualities of things.

I now want to suggest five reasons for a final resistance to the temptation to relegate secondary qualities to subjective status, reasons in part also derived from a naturalistic view of perception. That is to say, in what follows I am assuming the truth of the theory of evolution and its presuppositions, so I am not attempting circularly to justify my beliefs. What I hope to show is that when we take the theory together with what it presupposes about organisms and their perceptions, it does not support the view that the theories of physics are objective in a way which would deny objectivity to the deliverances of common sense.

We need first to consider the view of science which supports the illusoriness of secondary qualities. Obviously this comes down, roughly speaking, to the view that science and science alone gives us objectivity. And part of the reason for this is undoubtedly the fact— for it is a fact—that physics and the other sciences aim at an observer-independent account of the world. That is to say, in physics and the other sciences, the aim is to produce accounts of things restricted to their abstracted shape, size, number, position, and 'quantity of motion', in Galileo's words, and discounting their appearances to us. The thought is that the abstracted primary properties would be available to observers of any constitution, and would not depend on the specific make-up of particular observers.

However, from the fact that post-Galilean science aims to give an account of the world which abstracts from aspects of our or any sensibility—and is, in that sense, 'observer-independent', describing the world as if there were no humans or perhaps no observers at all in it—

it does not follow that what science actually does is not conditioned by our mental and physiological make-up. Science depends in all sorts of ways on our physiology, our mental make-up, and our activity, and on this point Rescher is right, even if he takes evolutionary considerations to be more restrictively linked to survival than they are. But science does not simply read the book of nature. What it discovers, it discovers as a result of our probing into nature, and in the terms in which we probe. Science, as much as everyday perception, is the upshot of human interaction with nature.

In the first place, there is a general constraint of scientific theorizing, arising from the need to test theories by observation. Where theories cannot be tested by us—and ultimately by our sensory faculties—they remain in the realm of metaphysical speculation. (This, of course, is the key difference between Greek atomism and modern scientific atomism.) To that extent, too, Rescher is right. Despite all the talk about science involving an observer-independent viewpoint, or one which is available to all observers, we have to entertain the possibility that a science confirmable by us has no overlap with one available to beings sensorily and intellectually quite different. Of course, we would not be able to *understand* their science; this indeed would be a consequence of a lack of a common empirical basis. But it is hard to see how a priori such a possibility could be ruled out. Davidson's well-known arguments against alternative conceptual schemes show at most that we could not translate or understand alternatives not locked into ours at certain key points, presumably at the most basic observational and inferential levels.[3] But they do not show that there could not be accounts of parts of the world we could not lock on to, or that, in some sense, such accounts might not be rational and articulated, which brings us on to our second point.

Secondly, then, the concepts which we apply to nature have to be expressed in terms which are intelligible to us, often in terms of models in which aspects of our everyday world are metaphorically applied to what underlies the everyday. It is by no means impossible that parts of the physical world may be impenetrable by our minds, constricted as they are by the world in which we live, think, and act. Bohr suspected that this was the case with the quantum world, and quantum theory since his time has done little to show that he was wrong. It may, then, be that if there were creatures capable of existing in conditions very remote from those of our empirical world on Earth, their per-

[3] Cf. D. Davidson, 'On the Very Idea of a Conceptual Scheme', in his *Inquiries into Truth and Interpretation* (Clarendon Press, Oxford, 1984).

ceptions and concepts would be so remote from ours as to be untranslatable to such an extent that even their theoretical explanations made no sense to us.

Third, as Ian Hacking and Rom Harré have in different ways shown, a great deal of contemporary science is artefactual: that is to say its subject-matter consists of effects and phenomena which we, as investigators, create and manipulate, which did not previously exist in the way in which we investigate them and which require activity on the part of our senses or instruments for their detection under the concepts with which we detect them. The fact that much of what we discover in science we also create does not impugn the reality or objectivity of what is thus revealed. What it does is to cast doubt on a naïve view of science simply as reading off essences from what is there without us. The artefactual nature of science is, though, quite consistent with an evolutionary view of things, in which organisms do not simply adapt to their niches, but change and mould them in various ways.

Following on from our third point, we might ask further whether the world reveals anything of itself, unless probed by one type of apparatus or another, either sensory or technological. Is it the case, then, that all knowing is interactive, and not merely secondary quality knowing? If one is prepared to go down this route, then the primary–secondary quality distinction becomes a matter of degree; it is only one between forms of perception that are more or less widely available. There is, indeed, some plausibility in this suggestion: the colour-blind dog and the deaf snake lack perceptual abilities that we have, just as we lack their sense of smell and responsiveness to vibrations on the ground. But we can all come to similar judgements about certain invariant and apparently stable properties of things and space and position. What, though, we have to guard against is the idea that any delineations of space we all share could arise or be deployed in the absence of any sensory interactions with the world. Equally, we have to guard against the idea that what is stable and what we can all agree on (humans, Martians, bats) is, by virtue of that fact, the essence of the world, as opposed to something abstracted from reality in its full and multifarious concreteness, particularity, and instability. And then again, there can be no a priori presumption that there has to be an account of the world on which all observers, from whatever vantage point and with whatever sensory apparatus, can agree. In a way the problems associated with the interpretation of quantum mechanics serve simply to underline this point. The comparatively unproblematic parts of quantum theory are the empirical predictions at the level of our instruments and senses. It is when we try to explain those predictions that our concepts seem to fail. But if there were creatures who were sensorily adapted to that world, and who could probe into it with greater ease than us, is there any guarantee that we could make sense of what they thought or perceived? In other words, what the world reveals is never independent of the mode in which it is approached.

Finally, as both Nancy Cartwright and Bas Van Fraassen have argued in different ways, much fundamental and not so fundamental work in physics consists of applying idealized models to phenomena and bodging things one way or the other until a fit is secured. As Heidegger put the point: 'Every new phenomenon emerging within an area of science is refined to such a point that it fits into the normative objective coherence of the theory'.[4]

These five considerations taken together detract from the view of science as involving some disconnected, purely rational observer reading off the timeless and external essences of things; they do not, though, do anything to show that what science discovers is either untrue or unreal.

A Naturalistic View of Knowledge

The picture of scientific activity as revealing patterns and regularities in nature in response to and partially dependent on our actively embodied probing is, indeed, just what one would expect from an evolutionary view of science. An evolutionary view of scientific activity is one which sees science as a process by which we, as an embodied species, attempt to extend our perceptual and cognitive niche both by isolating certain properties of objects and by pushing our perceptual and cognitive resources into areas beyond what is initially natural for us. In so doing, we will uncover some new coincidences and convergences between our faculties, instruments, and thoughts and the world, and at times create new aspects of the world. Sometimes the world will fail to respond to the invitation, and the probing will result in no projectible regularities or systematically manipulable effects, and then other lines of thought and other types of probing will have to be tried, as, once more, evolution would predict.

It does not, of course, follow that because a particular model of, for example, gas molecules has been derived from our experience of elas-

---

[4] M. Heidegger, *The Question Concerning Technology*, (Harper and Row, New York, 1977), 169.

tic balls, or because the model does not precisely fit actual gas clouds, that what it is being applied to is not real or that its predictions are not (more or less) true or that the elastic ball analogy does not reveal something real about the gas molecules. Nor does it follow that an effect we produce and which did not exist before we produced it, such as the photo-electric effect, is not real. Neither do the general restrictions on our theories due to what we can observe and conceptualize impugn their truth or the reality of what they are applied to. Indeed, as just remarked, the world does, in evolutionary fashion, select some theories and kill off others in which there is no predictive profit.

My point, though, is that science—despite attempting to produce accounts of the world which prescind from our sensibility and succeeding in so doing up to a point—nevertheless remains a human activity. Its results and theories are expressed in human language, conceived and expressed in human thought patterns, and constrained by them, structured and applied through human actions, and, in the end, corroborated through human perceptions. All this is in a broad sense an evolutionary view: actual science is one means by which a given species explores its world.

Science, then, is not a passive mirroring of the world. Its theories are a function of human activity and human imagination working on the world. In this working the world sometimes corrects and destroys what we imagine, but at other times is itself corrected or changed to provide a more congenial habitat for our theories and practices—again as happens elsewhere in evolution. Ortega y Gasset exaggerates in saying that: 'what we call nature . . . is nothing but a conjunction of favourable and adverse conditions encountered by man . . . it consists exclusively in presenting facilities and difficulties . . . in respect to our aspirations'.[5] He exaggerates in saying that what we call nature is 'nothing but' and 'exclusively' thus and so. Even what we *call* nature includes things we cannot know, which elude our probing and which are outside our cognitive niche, and it is important for objectivity to remember this. But if Ortega had said that what *our theories of science* tell us is thus and so he would be pretty close to what I am arguing here.

If the complaint about the illusory nature of secondary qualities is that they arise only in human interaction in the world, and are conditioned by that, much the same could be said about the theories of science themselves. And if the complaint is that secondary qualities are

---

[5] Ortega y Gasset, *History as a System and Other Essays* (Norton, New York, 1962), 224.

shown to be secondary by science, we have to inquire as to the basis on which science is taken to be the touchstone of reality, but the temptation to do that will surely be weakened when it is appreciated that science itself ought to be viewed as one particular way we humans have of interacting with and manipulating nature, and not as something timeless, essential, and trans-human. It is true—and part of science's success—that if and when you are interested in causal regularities and what underlies them and is responsible for them, then you will accept the ontology of science as delineating what you should work with, abstracting from properties which are not causally basic. But there are ways other than the scientific and causal by which we reach out to the world and uncover aspects of it, one such being the perceptual, which, as we have argued, cannot be simplistically reduced to the causal. Our perceptions of secondary qualities may well have been initially embedded in us because of their pre-scientific causal utility, but now that we treat them as scientifically secondary, it does not mean that we should not continue to treat them as useful and objective in other contexts.

Indeed, far from being disabled by science from treating secondary qualities and the like as useful and objective, so long as we are living a full human life, we cannot but do so. It is hard to see how an embodied consciousness such as a human being could perceive the world except from a specific time and place, and with a particular sensibility. Although the qualities delivered by our senses—sight, touch, taste, smell, sound—are just those which are dubbed secondary from the scientific-causal viewpoint, they are also those which provide the mechanism by which we link with the world. They are the means by which things in the world affect and stimulate our minds, meshing with something in us to attract our attention in particular ways. So long as we are human and embodied, the world will be bound to appear to us replete with secondary qualities, and this must be the premiss of resistance to treat these qualities as illusory or unreal.

The fact that we cannot live in or interact with a world deprived of secondary properties is tantamount to saying that we cannot live in or by the scientific image. Such a conception of the world would destroy meaning and value, because our meanings and values, our art and our ethics are deeply intertwined with secondary qualities, such as colour and sound, and with sensory reactions such as pleasure or pain. It would also deprive us as human beings of any relationship to itself, for we connect to the scientific image only via the manifest image. The manifest image then remains fundamental to our being in the world.

Indeed, calling it an image is rather misleading. Better, our sensibility and what it reveals is fundamental to our interaction with the world; from this point of view, it is the scientific image which is secondary. It is dispensable in a way our initial, sensorily permeated manner of being and acting cannot be. To quote Sartre, while physics may possess an absolute truth in its own sphere, this 'makes no difference to that other absolute, which is the world of perception and *praxis*'.[6] Of course, once we recognize that the scientific picture is also a result of human practice, just as much as the manifest image, the temptation to displace the manifest image of the everyday—the 'folk' image—should diminish.

We can, though, speak of illusions in particular contexts. A straight stick appears bent in the water. The fish appears blue because of the way the light is in the water. The moon appears to follow the walker over the rooftops. The sun appears to be going down behind a mountain. In all these cases, it is clear with respect to what and in respect of what we are being deceived. It is a very different thing to say that a whole way of perceiving is mistaken. By comparison with the bent stick or the blue fish, to dismiss a whole way of perceiving as illusory really does seem to be a matter of decision, rather than anything forced on us by the facts or by getting a better look at how things really are.

I suspect that the tendency to denigrate the common-sense view of the world is increased because those who argue in this way do not fully appreciate the costs of doing so, either positively, or negatively. Positively the cost is, as already remarked, the extreme strangeness of the scientific world-view, closed space-time, quantum theory, and the rest. Negatively, we must not think that we can keep bits of lived-world picture: we are going to have to say not just that it is an illusion that the table is brown, Westminster Abbey is grey, Melanie's hair is black. The table, Westminster Abbey, and Melanie herself are just as illusory, just as much part of the manifest image as their colours, just as much excluded from the scientific image.

The temptation to opt for the scientific image is increased by the tendency we have to express the results of our knowledge-gathering activities propositionally, which in turn helps to conceal the fact that all our knowledge, scientific and non-scientific alike, arises in active interaction between us and the world. While it is true that inquiry does produce truths, and knowledge may thus be said to be a product, there is also a process involved, something which should not be overlooked

6 J.-P. Sartre, *Cahiers pour une Morale* (Gallimard, Paris, 1983), 529.

in a naturalistic and evolutionary account of knowledge. Examination of the process will tell us something of the status of the product, which may well be forgotten if the product is viewed in abstraction. One thing which is likely to be overlooked is the point mentioned earlier, in connection with Bohr, that our knowledge cannot entirely outstrip our embodiment and experience. Nor indeed can that of any other creatures. So I am dubious about the claim made by D. H. Mellor that lacking certain experiences is not a *cognitive* deprivation.[7] If we take the naturalistic view I have been advancing, lacking a given type of experience will mean that certain aspects of the world will not be revealed to one. It makes a considerable difference to what one knows and to what one will be able to go on to say about (e.g.) colour whether one has colour vision or not—a point actually underlined by Hardin's denial that there is any determinate microstructural property shared by all objects which are experienced as blue. Someone lacking experience of blue will be unable to project 'blue' in the same way as a normally sighted person without studying the physiology and psychology of those who can see blue. My point here is that we should think of experience not as an ineffable quantum of feeling private to an individual, but as that which for a given species motivates and grounds particular ways of classifying things and proceeding in the world. Outsiders may be able to latch on to an experientially based concept in a rough and ready way, but crucially they will not be able to see the point of the classification in question. In this respect, as in others, there are parallels between secondary qualities and moral evaluations. In both cases one should be wary of concluding that we are dealing with projections; in both cases what we are talking about belongs to our primary and fundamental modes of being in the world and cannot be treated as projective—except by invoking theoretical considerations which we cannot live by.

Regarding knowledge as the result of a biologically grounded process should also help to resist the tendency to denigrate experiential knowledge and to assume that there can be knowledge which is not in some sense experiential. Once that assumption is made, it is easy to see why the equation is made between the more abstract and more

---

[7] Cf. D. H. Mellor, 'Nothing Like Experience', *Proceedings of the Aristotelian Society*, 92 (1993), 1–15. For a sustained and convincing rebuttal of the view that knowledge is basically propositional, see Bryan Magee's contributions to his joint correspondence with Martin Milligan, published as *On Blindness* (Oxford University Press, 1995), especially Letter Seven. Magee is fighting against philosophical fashion, but what he says will make sense to anyone who appreciates wine or music or who believes that it was only through certain experiences that he came to know the meaning of fear or of love.

general and the more objective. But what is overlooked here is the extent to which even the more abstract and general descriptions and modes of classification are produced interactively, in embodied inter-action between species and environment.

The position I am sketching here, which emphasizes the interest and activities within which knowledge is produced, may look like a form of idealism or anti-realism. I want, though, to resist this characteriza-tion, precisely because my perspective is a naturalistic one. While I have been emphasizing the contribution the knowing organism makes to its theories, classifications, and activities, the other side of the equa-tion is the controlling effect of the environment, natural, cognitive, and social. This controlling effect is by no means total, as I argued earl-ier, but it is there nevertheless, at the margin. Some ideas and modes of activity carry within them the seeds of their own destruction. Other ideas and modes lead to the destruction of those who hold them. The environment just will not wear them or their holders so long as they are wedded to them. Species get rooted out. In the case of ideas, the cognitive cost of trying to maintain, say, a geocentric view of the uni-verse is just too high. The cognitive and practical profit of maintaining an astrological perspective is not high enough (roughly nil). Base met-als could not be transformed into gold, albeit that both the notions of base metals and of gold arose in response to human probings, and without such or similar probings would never have arisen. As Harré puts it: 'the occurrent properties of the world . . . can never become available to us independently of the apparatus we have the ingenuity and technical skill to construct nor to any other species either'.[8] In the social realm, egalitarianism as a social aim goes too much against the grain of human nature to be carried through without grotesque dis-tortions and suffering. The mind, then, does not create the world or human nature.

We are, as I say, embodied and *interactive* with the world, and tak-ing this fact as a premiss in our epistemology pulls us away from pure idealism (or anti-realism). On the other hand, pure or naïve realism plays down our embodiment too, tending to treat our knowledge as the result of spectating an already divided up and categorized world. It is odd that philosophically we have largely disposed of this sort of naïve realism at the level of everyday existence, only for the same atti-tude to re-emerge at the level of science, on which, if anything, the

---

[8] Rom Harré, 'Exploring the Human *Umwelt*', in R. Bhaskar (ed.), *Harré and Critics* (Blackwell, Oxford, 1990), 297–364, at 302.

contribution of the mind and practice to both theory and classification should be clearer.

It is, of course, correct to say that science is one point at which human beings encounter a 'hard' non-human reality. Or rather, it is the place in which we try to investigate the world as it goes on (and went on and will go on) without us. And for this task we have learned that secondary qualities can be ignored. Science has had considerable success in its task, in terms of which it is perfectly reasonable to say that Newtonian physics is to be preferred to Aristotelianism. But the success of science in its task has blinded us to the extent to which even our most basic scientific concepts may be orthogonal to the world. Consider the concepts of cause, essence, substance, and event. Deployment of such concepts has revealed an extraordinary amount about the hard physical world. They are not mere projections on our part. They are widely applicable within certain domains (or cognitive niches) and are integral to many statements which are objectively true. But if they are not purely projective, nor do they unproblematically and universally categorize the way the world is. This should be clear enough from microphysics, but it is also the case that in much day-to-day science at the macroscopic level we also simplify and ignore exceptions to maintain the clear lines of causal and essentialist theories (another evolutionary trade-off, if you like).

What I want to say about science is what I want to say about other cognitive forms which are successful in their environments: that it is an interactive process between organism and environment which, in the light of certain sets of interests, reveals aspects of the world. Where I depart from those who would call the manifest image secondary or illusory is that I do not want to say that the scientific image is, absolutely speaking, primary. The cost of downgrading the manifest image does too much violence to our form of life; and it is not necessary either, once the nature of scientific knowledge is properly understood. My basic point here is that there is no theory of the world, no knowledge which is not the result of interaction between organism and world, and which is not, in its classifications and conclusions, coloured by the interests and perspectives of the organism.

I have dwelt on the ways in which natural science is so coloured, because the temptation to think that science is objective in a way in which everyday beliefs are not often comes from forgetting this fact. But in stressing the interest-relatedness of science I do not want to fall into the opposite trap of denying its objectivity. Granted that objects of any sort emerge or present themselves only against the background of a set of interests, there is no reason to impugn the objects (and objectivity) of science, because its objects and objectivity are in this respect no different. I argued in the previous chapter that we cannot move from success, evolutionary or otherwise, to truth of belief or theory in any precise way. Considered merely as an account of survival, the theory of evolution is quite consistent with significant amounts of cognitive shortcomings in survivors, and this is my objection to the attempts of Lorenz and others to move straight from survival to cognitive success. But, taken more broadly, an evolutionary perspective is a naturalistic one, one which sees us and other species as living and operating in a real world, and as disclosing aspects of it. An evolutionary perspective on knowledge, then, will presuppose that our cognitive activity takes place in an objective world and as a response to it. From the point of view of the agent, at least, there must be a presumption of objectivity about his best considered and least dispensable cognitive efforts, for these make sense to him only as attempts to probe the real world. If this presumption attaches to our scientific theories, it must also attach to our common-sense perceptions, secondary qualities and all, because these perceptions are no less grounded in our interactions with the world and are certainly not dispensable to us in these interactions. As far as living our lives goes they are actually far less indispensable to us than the speculative world pictures afforded us by modern science. To put the point directly, while we cannot live without secondary qualities, we can live perfectly well without astrophysics.

My conclusion, then, is that the theory of evolution in the strict sense of an account of species survival and origin, cannot justify our basic epistemological standpoint against scepticism. Nor can the 'tournament of the mind' envisaged by Smithurst either explain our motivation in searching for truth or ground epistemologically the theories so discovered. Nevertheless, if we take the realist attitude to the world presupposed by evolutionary theory, and focusing on the interactive naturalism that theory suggests, we should equally be wary of drawing sceptical inferences from evolution—for natural science, which might seem to motivate such scepticism, turns out to be in significant ways no less interactive than the everyday secondary quality image it might seek to displace. It all depends on which interactions one is prepared to think of as disclosing real aspects of the world, for without interaction, nothing will be disclosed, a consideration which fits well with an evolutionary perspective which sees us and other organisms as perceiving through interaction with the world.

There seems no reason either in principle or in practice to think of the full-blooded image of everyday life as less real or less true than the rather more abstract and attenuated accounts given in modern science. The temptation to do so is, of course, a feature of our ability to reflect on and to question, as we saw in Chapter 2. And it is true that while we are embodied, and firmly rooted in the natural world, we are also reflective. We can, and should, in various ways transcend and refine the given. The question here is not so much the limits of what we can probe into. Whether we can get far with evolution, biology or astrophysics or quantum theory would seem to be largely an empirical question, depending as much on ingenuity, happenstance, and motivation, as on any a priori limits to what can be achieved by us as a species. The question is rather what we should do with such knowledge once we get it—how far we should allow it, and more directly philosophical reflection, to undermine our confidence in our basic orientation to the world.

In practice that orientation is secure, as Hume was the first to admit. We are real creatures in a real world, and nature will not allow us to forget it. I do not think that having forgotten or pretending in our reflectiveness to forget that, the theory of evolution can pull us back into the real world.

But if, for the sorts of reasons adduced here, we decline to take the step into forgetfulness, a naturalistic perspective can help to reinforce our confidence in the given, even at the level of reflection. What is often called evolutionary epistemology, like philosophical naturalism more generally, often seems content with accounts of how we have come to have various beliefs and perceptions, telling us that such and such has come about in the struggle for survival and reproduction. From this starting-point some would derive optimistic and others pessimistic conclusions, as we have seen. But epistemology should not avoid normative, justificatory questions. In this sense, it is an aspect of our self-conscious reflectiveness. What I hope to have done, particularly in this chapter, is to have suggested that a naturalistic starting-point can throw light on what we ought to believe so as not to undermine our natural confidence in the deliverances of everyday perception.

6

Morality and Politics

Self-consciousness makes us aware of the fact that we are separate from the world, and that what we believe about the world may not be true. Reflection on ourselves as embodied, and as having an evolutionary history, gives us some assurance of the efficacy of our beliefs, but it also makes us aware that efficacy is not the same as truth, and this is quite apart from the further aspect of self-consciousness about our beliefs, the realization that what has in the past been efficacious need not continue to be so. Epistemologically, evolution together with self-consciousness should produce in us a degree of caution, though tempered with the assurance that as an embodied species in a real world, our cognition is a possible if limited perspective on that world.

When we come to look at morality, we find that self-consciousness and evolution again pull in somewhat different directions, dramatically and at times in unexpected ways. To start with, talk of morality is itself ambiguous. Do we mean morality as that which is done and taught and enforced within a particular group? Or do we mean that which in some absolute sense simply ought to be done, regardless of group norms and loyalties? And what, if anything, is the connection between the two?

Evolution would seem to bear on morality in the following way. In general terms, evolution is about survival and reproduction, though whether of individual organisms, groups of organisms, or the genes of individuals is disputed and will make a difference. For members of a species such as ours, which both has certain vulnerabilities and which is capable of co-operation, some inherited tendency to co-operation will be advantageous. Darwin himself recognized this, attributing mankind's 'immense superiority' to other forms of life in part to 'his social habits, which leads him to aid and defend his fellows'.[1] But, how far an advantageous social habit amounts to morality in the full sense is precisely what will concern us throughout this chapter.

---

[1] C. Darwin, *The Descent of Man* (John Murray, London, 1898, 2 vols.), i. 72.

In self-conscious individuals, such as ourselves, the practice of co-operation will be enhanced by the belief or feeling that we ought to co-operate with other members of the group in question. So something like a sense of 'ought' can be seen as having evolutionary and even biological roots. If, as I shall argue, some tendency to co-operation is biologically advantageous, inherited impulses in us towards co-operation will be explicable, and to be expected. Biologically, it is not to be wondered at that we naturally have feelings of sympathy towards suffering fellow humans, as having such feelings is mutually advantageous. Sociologically, these feelings may be reinforced by cultural practices, encouraging such feelings and their practical expression in actually helping people in need. But, in a self-conscious species, as with cognitive belief, it is not going to be easy to confine or analyse a sense of ought to inherited or learned impulses or practices or to confine it to what is inherited or learned. Why should these givens define or confine what *ought* to be done? Moreover, there are strong reasons to think that the logic or dynamic of self-consciousness is towards a universal recognition and acceptance of what I do, and not just acceptance from some limited group or in so far as I happen to receive benefit from an impulse of sympathy from someone else.

Self-consciousness may, as Hegel thought, be tied up with gaze, with the recognition of me as a self by the other. I certainly look for acknowledgement by others that I am not just a thing, but self-conscious, and this imposes on me a reciprocal demand to acknowledge them. So, as with knowledge, while there are reasons for thinking that evolutionary advantages can arise from something like a co-operative or moral self-consciousness, self-consciousness quickly transcends the narrowly advantageous or merely felt in the direction of something a Kantian philosopher might recognize as moral: an unconditional duty of recognition from other rational beings owed to any rational being, whether part of that being's group or not (which parallels the similar epistemic ideal of a set of beliefs acceptable to any ideal observer, whether possessing any particular sensibility or not). We will need to assess the full-blown Kantian or Hegelian perspective, in which morality consists essentially in being a person and showing respect for other persons. But first, we need to look at the biological and sociological accounts of co-operation and altruism. Only then will we be in a position to see how far the notion of a person and the associated structure of morality might be based on our existence as biological and social beings and how far it transcends this existence.

Biological 'Altruism'

Altruism is, as we shall see, a misnomer when it refers to anything definable or expressible in biological terms. Altruism, in the strict sense, involves activity in which even if I do not actually lose, I do something for someone else, without regard for any actual or possible advantage to me. I do something for the other because I want to benefit him or her, and not for any other reason. In biology, we do not hear of behaviour which is both selected for in natural selection and not in the short or medium term for the good of the agent. I follow custom in using the rebarbative expression selected *for*, because it is possible that non-useful behaviour might be selected, as a by-product of some other behaviour or faculty that *was* useful, and which was, accordingly, selected *for* in natural selection. In the jargon, the non-useful habit would be selected, but not selected *for*; that is to say, it would not be *because* of its own character that it was selected. How, though, might natural selection favour (select for) a faculty or type of behaviour which was not useful, when the theory and its explanatory mechanisms are about selective and reproductive *advantage*? As one eminent Darwinian, Richard Dawkins, puts it, from the human point of view, the Darwinian world is very nasty indeed. Far from attempting to imitate it, we should do our best to run away from it. (It has to be said, though, that apart from disparaging Holy Books which recommend just such a course of action, this particular Darwinian is quite unable to explain why we have an obligation to act against our 'selfish' genes.)

Saying that evolution is about selective and reproductive advantage does not, of course, mean that co-operative behaviour is ruled out. On the contrary, in conditions of scarcity and individual weakness in the face of a hostile environment, co-operation is very much ruled in. In many situations, real and simulated, the pursuit of short-term individual self-interest will leave individuals worse off than if they co-operate. An illustration of this is given by the prisoner's dilemma, an essay in game theory in which the fate of two individuals is inextricably linked. Two prisoners, A and B, are both accused of a crime and held and questioned in isolation. If A denounces B, and B does not denounce A, A will get one year in prison and B ten, and the reverse will be the case if B denounces A without A denouncing B. But if B denounces A, with A also denouncing B, both will get six years. If both stay silent, then both will get two years.

If you and I are in a prisoner's dilemma, then what is in effect cheating on my part will produce the best outcome for me, but only so long as you play the game. If you cheat as well (hoping to maximize your advantage), then we both lose more than had we both played the game fairly. The worst outcome for me is if I play the game and you cheat (and you get your best outcome). Mutual co-operation, by contrast, produces a satisfactory result for us both, in which neither of us gains as much as we might have had either of us been playing against a 'sucker' (someone who plays by the rules against a rule breaker), but we both gain more than had we been playing against a cheat.

The problem, of course, is to know whether one's opponent is going to play fairly or not. Let us assume that evolution or convention has given members of a group a disposition to co-operate, as being the most advantageous strategy overall, that is the one least likely to lead to disaster, for individuals, as well as for groups. In such circumstances, where in effect co-operation could be generally relied on, wouldn't an individual who cheated, correctly expecting the rest to play the game, do best of all? And so wouldn't individuals of that type become evolutionarily dominant, and eventually through their equally dominant offspring upset the genetically based strategy of co-operation?

The answer is that it would, were all cheating interactions regarded in isolation. But if the interactions are repeated, the co-operators might well evolve a system of punishing cheats, by not co-operating with them on succeeding occasions. It can, indeed, be shown that a tit-for-tat strategy can uphold a system of co-operation and prevent its subversion by cheats. Not only will tit-for-tat sustain an existing system of co-operation. In repeated prisoner's dilemmas, initial non-co-operation of agents who have not yet set up conventions of co-operation can be punished and eliminated by repeated exercises of tit-for-tat. So from the behaviour of agents who are intent only on their self-interest, a system of what looks like co-operation and associated punishment for cheaters can evolve, and on occasion has apparently done so. The not unreasonable premiss underlying this type of development is that over the long term universal or near-universal abiding by the rules is in the individual interest of all involved. Short-term losses through co-operation turn out to be equivalent to long-term prudence for an individual provided that the rules can be made to stick for all. In *The Ant and the Peacock*, Helena Cronin gives examples of tree swallows, baboons, mongooses, and other creatures who reward co-operation and punish defection as the prisoner's dilemma supplemented by tit-for-tat would predict.[2]

So evolution can produce what looks like genetically based co-operation. But the prisoner's dilemma plus tit-for-tat is very far from genuine altruism, or from morality at all. The co-operative strategies in question are clearly for the mutual self-interest of the parties involved. What game theory and the evolutionary examples demonstrate is not that self-interest is overcome in these cases by some co-operative urge, but rather that strategies of co-operation (and underlying urges and even genes) are, in contrast to first appearances, not inconsistent with self-interest. What is shown is that self-interest on some occasions actually requires co-operation, but not on all occasions. If there is no chance of a punishing tit for my self-interested or cheating tat, then self-interest and evolution will favour self-interested self-aggrandizement at the expense of others. If, as moralists are, we are interested in why we should feed the starving on the other side of the world, we will get no help from tit-for-tat prisoner's dilemmas. Or rather, the only help we will get will be through the devising of implausible scenarios in which our survival depends in some way on the starving—thus inadvertently encouraging strategies of international terrorism on the part of the worst-off to ensure that our security is related to our doing enough to make them desist from such strategies.

Not only is the dilemma not engaged where reciprocal rewards and punishments might not be at issue, but the dilemma tells us nothing about what we *ought* to do as opposed to what would best serve our interests. The games of prisoner's dilemma and tit for tat in no way fill out the thought that we are doing all this out of respect for our fellow prisoner. The assumption indeed is that we would cheat on him if we could get away with it, and that he would do the same to us. That theorists talk in such cases of *reciprocal* altruism ought indeed to give us pause, for reciprocal altruism is something the parties engage in for joint individual benefit, and is not altruism properly speaking at all.

Reciprocal altruists are not acting against their best or long-term benefits, and are no exception to the Darwinian principle that there is no instinct injurious to its possessor, 'for natural selection acts solely by and for the good of each'.[3] What, then, of those cases, especially in the insect kingdom, where we see what looks like genuinely unselfish co-operation, sometimes even unto death, between members of the same species? The case is particularly striking where the suicidal

---

[2] Cf. H. Cronin, *The Ant and the Peacock* (Cambridge University Press, 1993), 258–9.
[3] C. Darwin, *The Origin of Species* (Penguin, Harmondsworth, 1982), 229.

co-operators are sterile, and work and die assisting and defending the fertile members of the tribe.

The sociality of insects and other apparently altruistic phenomena like the birds and mammals who put themselves in danger by giving alarm calls to the rest of their group who then hide or escape, were at one time explained in terms of group selection. Some later analyses of 'alarm call' behaviour have questioned whether such behaviour is really individually disadvantageous; after all, the bird who gives the alarm gets all his flock to show themselves as well! But certainly some types of apparently widespread and instinctive behaviour are seriously and indisputably disadvantageous to their individual possessors. Could the individuals be acting on behalf of the species or group to which they belonged? On this type of analysis, the species becomes the agent, not the Darwinian individual.

For a considerable period since the time of Darwin himself, evolutionary theorists of various sorts attempted to explain self-sacrificial behaviour on the part of individuals in terms of group selection. In other words, the unit of selection—that for whose sake the behaviour was done—was taken to be not the individuals but the group to which the individuals in question belonged, and for whom the individuals sacrificed themselves. It was never properly explained how the selection of groups was supposed to work on the behaviour and genetic structure of the individuals in the groups. Unless the individuals in the group actually benefited from being in the group, and did not simply sacrifice themselves for the group before reproducing, it is hard to see how any gene which encouraged self-sacrifice and personal disadvantage would not die out fairly quickly. Nevertheless, the producing of the self-sacrificial individuals was taken to be an adaptation for the benefit of the group, the species, or sub-species to which the individuals belonged. But how adaptations beneficial to a group are supposed to have an effect on the genes individuals transmit to descendants, if those adaptations require individuals to sacrifice themselves before they reproduce and pass anything on to their descendants is left quite obscure. This difficulty is compounded once we realize that in a population consisting of self-sacrificial individuals, those selfish individuals who benefit from the self-sacrifice of others will quickly outbreed and come to dominate the rest, and so the balance of selfishness and self-sacrifice which would be good for the group as a whole will be lost.

If group selection is a problematic notion in biology, particularly when one considers the genetic mechanics which would have to underlie it, individual self-sacrificial behaviour certainly does exist. How can that be explained on the assumption that individuals are seeking their own good, and that only individual behaviour seeking the reproductive good of individuals can be explained in evolutionary terms? The answer is that it cannot, and that its prevalence must be a real problem for evolutionary theory so long as we see evolution as working for the good of whole individuals.

If, however, we regard the unit of selection not as the whole individual and even less the species or group, but the gene or a cluster of genes, the evolutionary problem of altruism can be defused. An individual sacrificing himself or herself for a closely related individual will be helping his or her genes, in so far as kin share genes. To use the current biological terminology, whole individual organisms are to be regarded as the vehicles of replicators, with genes themselves as the replicators. The survival and reproduction which evolution is interested in is the survival and reproduction of genes. This being so, we can account for behaviour which looks like the effect of group selection, but which is impossible to explain in such terms.

The view that apparent altruism towards my kin is really selfishness towards those of my genes which are present in the kin I help would certainly go some way to explaining the possibly self-sacrificial alarm calls given by birds and animals to their kin and more emphatically, the efforts of sterile insects on behalf of their fertile relatives. In both cases the apparently altruistic co-operator is helping his or her genes in so far as they are in those he helps, while in the case of the bird giving the alarm call, at least, he would expect to be helped himself in a similar situation were the roles reversed, and his kin member in danger had the same genes as him and warned him. Interestingly, in some cases of self-sacrificial alarm-giving, it seems that alarm givers are more ready to give alarm calls to their close kin than to other members of the same species. Social insects in a particular nest or hive are, of course, all very closely related (except for slave ants captured from other nests who are duped into toiling for non-kin). Moreover, the female workers among the hymenoptera who labour on behalf of their mother's fertile daughters are far closer genetically to their sisters than are their brothers who do not so labour.

Kin-selection, then, explains the appearance of altruism away and, as such, may seem to throw little light on altruism or even self-sacrificial honesty towards those to whom we are not related—as Kantian morality demands. While this is true, kin-selection, along with reciprocal altruism, does show that *co-operative* behaviour can have an evolutionary base, even if it does so by underlining the fact that co-operation need not involve altruism properly speaking.

While the selfish gene analysis may well be true in the case of the social insects, who are usually genetically close to those around them, it has been argued that there is a form of altruism which could for a time be an evolutionary stable strategy within a group without needing explanation either in terms of reciprocal altruism or the terms of genetic proximity of those involved. By speaking of an evolutionarily stable strategy, what is meant is that the population would continue to consist of similar proportions of altruists and cheats, without being overwhelmed by ever greater proportions of cheats. This would be in a type of group in which altruists were relatively disadvantaged in comparison to cheats, but in which the greater the frequency of altruists, the greater the absolute fitness of altruists and cheats alike.

The point about such a case (which is discussed by Elliott Sober)[4] is that the altruists are not really suffering for their altruism, because they benefit, along with everyone else. Too much cheating by the cheats or too many cheats will reduce absolute fitness of individuals in both groups. It is, though, hard to see how there will not be a continual drive to increase the number of cheats in the population, and that, as Sober admits, stable proportions of altruists and cheats in the population as a whole will depend on excessive cheating and the tendency to the elimination of altruism being confined to sub-groups of the population, which either go extinct or are colonized by influxes of altruists from outside.

If excessive cheating leads to the downfall of a group, and the defeat by groups consisting of co-operators—as indeed it may—we can see that the type of group to which an individual belongs can have an effect on his or her survival.[5] But while, in this type of case, group membership has a causal effect on which individuals survive, it is not the group which is being selected for any more than in the selfish gene analysis the individual organism is being selected for itself. In both cases what is being selected is an environment which promotes the interests of participants, its individual members, the individual organisms, and ultimately the genes within organisms. So while it would be wrong to rule out talk of groups in selection altogether, for being in a group can have a causal effect on the survival chance of individuals, introduction of such talk does not necessitate the implausible reintro-

    [4] Cf. E. Sober, *The Nature of Selection* (MIT Press, Cambridge, Mass., 1985), 328 ff.
    [5] Cf. Kim Sterelny, 'Understanding Life: Recent Work in Philosophy of Biology', *British Journal for the Philosophy of Science*, 46 (1995), 155–83, at 170.

duction of groups as selection units. Groups are evolutionarily relevant to the extent that they increase the fitness of their individual members. What we would have in such groups would be various versions of the reciprocal altruism of prisoner's dilemmas plus tit-for-tat, rather than genuinely self-sacrificial co-operation. And even if the prevalence of drives towards reciprocal altruism sometimes led to genuinely self-sacrificial behaviour, this could, in the nature of the case, be only the exception. The genuine self-sacrificers would not breed and a population in which such a tendency was marked would be highly vulnerable to a take-over from within or without from free-riding mutants ready to exploit the potentiality for self-sacrifice. The population would not, in other words, be evolutionarily stable, and its downfall would be predicted by evolutionary theory. As Darwin himself put it, in social animals natural selection will adapt the structure of each individual for the benefit of the community, but only 'if each in consequence profits by the selected change'.[6] The inflexible law of the destruction of injurious variations would otherwise apply. The trick in considering group effects is to focus on *which* individuals are being selected (genes, rather than whole organisms) and to realize that in producing advantages for individual genes, there is more than one way of killing the cat.

More generally, there can obviously be evolutionary selection of self-restraining behaviour, where the over-zealous or over-aggressive tend to die prematurely through being killed or through consuming all their food source or in some other way undermining it. A possible (though disputed) example of the latter is the myxoma virus, which preys on rabbits. Over-virulent and hence highly reproductive myxoma viruses kill their host rabbits too quickly (and so die themselves) before they can be picked up and passed on to other rabbits by mosquitoes (who bite only live rabbits). So natural selection could certainly favour a degree of avirulence in this case, even though avirulence is, on the face of it, selectively less advantageous. But what is being selected here is not altruism, but prudence, and prudence, being self-regarding rather than other-regarding, is not the same as morality, let alone altruism.

While prudence is clearly not excluded by natural selection, it may also be that some kind of co-operation more extensive than either reciprocal altruism or kin-selection is not excluded either. It is certainly possible—as Egbert Leigh has argued—that there could be cases where what is good for a group is good for individuals too. Where this

[6] *The Origin of Species*, 133.

is the case we would not get the short-term advantage of individuals acting against the long-term advantage of the group as a whole. As Dawkins comments, in such a case 'apparently unselfish behaviour' could evolve, but not because individuals sacrifice themselves. It would rather be because 'individuals are not *asked* to sacrifice their own welfare'.[7]

Morality and Rationality

In morality, though, each of us is sometimes asked to sacrifice our own welfare, and that of our kin. It is this—and its arguably enervating effect—that enrages critics of morality from Thrasymachus to Nietzsche to Bernard Williams.

Nevertheless, given certain genetic dispositions to co-operation, albeit of a limited sort, given consciousness, and given reason (or self-consciousness), morality has to evolve, even if it puts a premium on behaviour which—contrary to evolution—militates against individual self-interest, either my own or that of my genes.

In saying that morality has to evolve given certain conditions, what I mean is this. We have already seen that reflectiveness about knowledge pushes us in the direction of drawing a distinction between beliefs which we simply have and beliefs which are true, independently of our holding them. We are also led to distinguish between beliefs which are useful and those which are true. Transposed to the area of practice, we would then be led to distinguish between *various* practices, the practices we *happen* to have, and practices which are desirable in a wider sense, and between practices which are useful and those which are good, apart from their being useful to particular individuals or groups, even if they are useful. This point needs careful stating. What I am driving at here is in the first instance a version of Moore's open question argument, or a version of the Socratic point we looked at in Chapter 1.[8] As reflective beings we are driven to ask questions about which courses of action are for the best, a question which cannot simply be stopped by the reply that such and such a course is conducive to satisfying my present desire. Why is it good, or for the best, to fulfil my *present* desire as opposed to my future desire? Why

---

[7] Richard Dawkins, *The Blind Watchmaker* (Longman, Harlow, 1986), 268; cf. Egbert Leigh 'How Does Selection Reconcile Individual Advantage with the Good of the Group?', *Proceedings of the National Academy of Sciences*, 74 (1977), 4542–6.

[8] Cf. Plato, *Phaedo*, sect. 99.

should it be best to satisfy a present desire I have, or even a set of future desires of mine? At this stage, it would still be possible to reflect and to decide that my present (or future) desires are best, and while this opens a gap between my desires and the good, it closes the gap, as I shall now argue, all too quickly. For reflection on what is good sees each one of us as members of a community of rational inquirers, in terms of which naked self-interest cannot count as a fully justifying reason where one person's interest cuts across the interests of others. (The cynicism of a Mandeville, far from being a refutation of this point, is actually an endorsement of it. While Mandeville argues for the pursuit of self-interest, he does so because he believes, or professes to believe, that pursuit of self-interest actually produces the best results all round; and while we may suspect this particular author of cynicism in so arguing, the position is not *per se* absurd or self-contradictory.)

Why is it that reflective self-consciousness on its own might take us from what is, in effect, prudence to something more like the ethical? As a first step, reflectiveness will certainly induce us to distinguish within our own desires and practices between those which are better and those which are worse, and between those which are merely means to other ends and those which are regarded by us as ends in themselves. So reflectiveness will lead us to rank our various desires and practices, and to make distinctions of quality between them. This in itself, though, may not give me reason for putting a value on other people's desires. It may not get me from egocentric prudence to an ethical state of mind, in which I regard the interests of others as having a value (and, at times, a paramount value).

What we need to realize in the practical sphere, as much as in the cognitive sphere, is that a reason is a reason, and, as such universalizable, or justifiable. That is to say, when we speak about reasoning (reflecting) about what ought to be done, just as when we start reasoning about what is so, we start to transcend our immediate desires and feelings, and look for beliefs and practices which can be justified or argued for. It is, of course, the most basic constraint on justification and argument, that the conclusion of a process of justification and argument should be universally acceptable. As we saw in Chapter 3 in considering language, language itself is essentially social in enabling us to reason through its symbolic and inferential resources. At the same time it opens up both the possibility and the need for each one of us to have his or her beliefs and reasons endorsed or criticized by other speakers. Reasoning thus makes at least an implicit reference to a community of ideal reasoners, within which one's conclusions will be endorsed.

So far, none of this militates against purely prudential considerations regarding practice. And indeed, I can surely look at the behaviour of a consistently selfish and self-regarding individual, and say that he has good reasons for doing what he does. I can see that, given his aims, he is rational. But can I say the same about his aims? I think that the answer to this question all depends on the circumstances.

Looked at from outside, I can understand the rationality of a consistently selfish individual, one who simply maximizes his own interests. I may also understand and accept many of his goals as providing good reasons for action—for example, avoidance of pain, search for honour, refinement of aesthetic taste, cultivation of intellect. Many of these aims I share myself, and those I do not, I can still see as designating goods in a man's life. So, looked at from outside and so far, reason can find nothing to quarrel with. The man is behaving rationally in adopting the aims he does, and in pursuing them in the way he does. And, were we all self-sufficient individuals in a state of plenty, reason and reflectiveness might have little more to say on the matter of conduct.

But, of course, we are neither self-sufficient, nor in a state of plenty. My perfectly rationally acceptable goals cannot be fulfilled without at times preventing you from furthering your equally rationally acceptable goals. The ethical realm arises in a context of mutual conflict, where the reflective agent asks himself how best to resolve the conflict. One way, of course—the way of Thrasymachus in *The Republic* or of Nietzsche in *The Will to Power*—would be to let the stronger win every time, but given our minimal constraint on rationality this would not, without further argument, be rationally acceptable. At the very least the physically or intellectually weaker participant in the debate will justifiably want to know why his reason for avoiding pain or hunger, say, is being discounted in favour of my avoidance of pain or my pursuit of honour. My *desire* to avoid pain or to pursue honour is being satisfied, but is it *reasonable* that it should do so at the expense of someone else? To show that it is rationally acceptable, it will need to be shown to be acceptable to others not involved and even to the weaker person. I am not saying that this could not be done, or indeed saying anything about what reasons of this sort might look like, but pointing out only that what is rationally acceptable is not to be analysed purely and without remainder in terms of what I (or any other individual) desire.

Bernard Williams has argued against this sort of argument from rationality to ethics on the grounds that the rationally reflecting I is still me, with my concrete beliefs and desires, and that I am not 'required to take the result of anyone else's properly conducted deliberation as a datum, nor be committed from the outset to a harmony of everyone's deliberations'.[9] He distinguishes between the two things I am here trying to align: between the *impartial* I of reflective truth-seeking and the reflective but still partial I of practical deliberation.

An initial comment on what Williams says would be to say that there is indeed nothing to stop the I of practical deliberation from riding roughshod over the desire of others or from ignoring them altogether. The question is whether he would be acting unreasonably in so doing, and this in turn depends on one's conception of rationality. Williams says that an I which takes on the perspective of impartiality will lack the identity to live a life that respects its own interests. While this is true of a completely depersonalized impartiality in which I abstract from any concern particular to my own life it is not clear that any generally acceptable form of practical reasoning requires anything like complete impartiality in that sense, or indeed any assumption of formal equality of interests. One might regard the best situation overall as one in which people were to be free to pursue their own desires and lives subject only to a minimum of respect for the rights of others, and in some circumstances one could argue for this quite rationally, reasonably hoping for agreement on this even from those at the bottom of the pile. What, *pace* Williams, it seems hard to envisage is for someone to say that he is simply going to ignore the desires and views of others and pursue his own to the exclusion of any others, and still be regarded as reasonable. This would seem to be the paradigm case of unreasonable behaviour. One might suggest that such a man would have been better off not arguing or discussing at all, but that too seems tantamount to his arbitrarily closing off his powers of reflection.

Williams's decoupling of rationality and ethics is in fact part of a deep-running pluralism about morality. As he says 'value-conflict . . . is something necessarily involved in human values . . . where (such) conflict needs to be overcome, this "need" is not of a purely logical character, nor a requirement of pure rationality, but rather a kind of social or personal need, the pressure of which will be felt in some historical circumstances rather than others'.[10] Renford Bambrough, who draws attention to this and cognate passages in Williams suggests that part of Williams's underlying rationale here is a 'milk and water positivism', which can find nothing 'in the world' which value judgements,

[9] B. Williams, *Ethics and the Limits of Philosophy* (Fontana, London, 1989), 69.
[10] B. Williams, *Moral Luck* (Cambridge University Press, 1981), 72.

as opposed to factual statements, are to fit, and so finds he has no use for any purely logical demand for consistency in the realm of values.[11] Yet, as Bambrough urges, is there not a requirement in reason for consistency in areas other than the empirical? Is there not something wrong if I have a set of conflicting intentions, or if two people offer me advice, but do not agree in the advice they offer? Is there not something wrong if Williams's philosophical thoughts conflict with mine, albeit there is nothing 'in the world' for them to fit? My practical choices are, indeed, *mine*, and the individual agent is autonomous and a judge of good and evil, but analogous things are true of my empirical judgements as well. Furthermore, the *mineness* of my choices and value judgements no more stops relations of logic holding between them and between yours than it does between my philosophy and yours or between my empirical judgements and yours; nor does it render *merely* psychological the quest for logical harmony between my choices and yours.

So far we have been considering self-consciousness in purely argumentative terms, as involving a certain degree of generality and impartiality in arguments. When, though, we look at self-consciousness phenomenologically the need for mutual recognition is underlined and reinforced.

For Aristotle, the man who is self-sufficient, who has no need of others is either a beast or a god. This is partly because of our lack of self-sufficiency and need to live in structured co-operative groups, the family, the village, and the state. But it is also because of reason itself, which, perhaps following Plato, Aristotle sees as essentially dialogic. Speech, says Aristotle, is different from voice, which is possessed by other animals, as well as by us, allowing them and us to express pleasure and pain. By contrast, speech serves to indicate what is harmful and what is useful, 'and so also what is just and what is unjust'.[12] It is our perception of good and evil, of just and unjust, which distinguishes us from animals and it is the sharing of a view on these matters which makes a household and a state for Aristotle. The stress on a common view in politics is because in perceiving the just and the unjust, and the good and the evil, we are perceiving things for which *reasons* have to be adduced and which take us beyond purely natural causation, and because reasoning presupposes potential communities of inquirers capable of recognizing each other's reasons as such, from

[11] R. Bambrough, 'Ethics and the Limits of Consistency', *Proceedings of the Aristotelian Society*, 90 (1989), 1–15.
[12] Aristotle, *Politics*, 1253a7.

an impartial point of view. It is, indeed significant that Aristotle has to argue that those he wishes to keep out of politics (women, barbarians, slaves) are naturally deficient in the full possession of reason. That our modern sensibility has no stomach for the exclusions Aristotle argues are natural is actually a backhanded endorsement of his view of reason as essentially interpersonal—though we are naturally rather upset when some of those lately admitted into the body politic turn out themselves to have views on strangers closer to Aristotle's than to what is politically acceptable in polite society.

Self-Consciousness and the Other

Nevertheless, following Hegel we can, I think, expand and deepen the Aristotelian insight that self-conscious rationality itself subtends a notion of practical reason in which acceptance by others of what I do is essential to the acceptability even for me of what I do.

To do this, I want first to consider some biological speculation about the origins and usefulness of empathy or fellow-feeling in the animal kingdom. The claim, associated particularly with the work of Nicholas Humphrey,[13] is that for creatures living in complex social groups, social interaction would be greatly helped if those creatures have some awareness of how their fellows will act and how they feel. Awareness of this sort would not be an issue for creatures who act rigidly, according to pre-set patterns of response, as some claim is the case with bees. But where behaviour is flexible and plastic, and has an air of responding creatively to new situations, it would be greatly to the advantage of members of a group to be able to empathize with their fellows, whether these are colleagues or rivals. Empathy in this sense would involve sensitivity to the feelings and projects of the others, having an inchoate grasp of their inner life, as it were. And if a creature can empathize with the feelings of others, why can it not have some sense—vague and inarticulate, no doubt—of its own inner life?

Having an inner life, then, and being sensitive to those of others is of evolutionary advantage. Knowing what others will do and think in given circumstances is the basis of deception. In deception I do or express something not in order to achieve the normally intended result, but in order to get a fellow creature to act as if I had done that thing or had expressed myself truthfully. There is, indeed, evidence

[13] Cf. N. K. Humphrey, 'The Social Function of Intellect', in P. P. G. Bateson and R. A. Hinde (eds.), *Growing Points in Ethology* (Cambridge University Press, 1976), 303–18.

that some higher primates, such as baboons and chimps, do engage in deceptions on their fellows, apparently able to respond to what their fellows might make of their actions.[14] One chimp, for example, has been observed suppressing a grin of fear, a sign of nervousness, when challenged by a stronger rival. A plausible explanation of this type of behaviour would be that the weaker chimp is sensitive to how he would react were he the stronger one, observing signs of fear in his rival. In other cases, baboons have been observed uttering alarm signals with the effect of scaring rival baboons away from food, and possibly with the intention of securing the effect. The precondition of such subterfuges, if such they be, is that at some level the deceiver is able to imagine himself in the shoes of his rival, knowing how the rival would react in specific circumstances. Social interaction, co-operative or competitive, will be greatly helped by my being able to anticipate the reactions of others and, in that sense, feel myself in their shoes, and this ability to empathize with others will clearly have evolutionary advantages. The ability to see both oneself and others as having an inner life does indeed form part of what is required for the full possession of self-consciousness which comes with membership of a linguistic community and will doubtless play its part in the tournament of the mind which we referred to at the end of Chapter 4. The ability to think explicitly about one's own state of mind and about those of one's fellows could well be founded in a quasi-instinctive empathy at a lower level of evolutionary development; such empathy would naturally help to provide the content for such thinking, when we reflect that another is in pain or angry and so on.

These speculations about the evolutionary background of self-consciousness are unsurprisingly analogous to the views of Davidson and Dennett on the nature of intentional language. In the view of

[14] For an up-to-date and comprehensive discussion of both sympathy and deception among primates, see Frans de Waal, *Good Natured* (Harvard University Press, 1996), esp. ch. 2. The upshot of de Waal's descriptions seems to be that higher apes can and do empathize with each other's emotional states, act on such empathy, and may do things which mislead their fellows so as to gain advantages over them. It is unclear that the behaviour which leads one to those conclusions supports the claim that the chimpanzees and other primates in question consciously attributed to each other intentional states (states of belief, knowledge, emotions, etc.); indeed, in the absence of language it is hard to see how there could be such evidence. This is relevant to the general topic of this chapter, in so far as morality proper requires not just a feeling for others but a recognition that they, like me, are intentional agents, choosing what they do and so open to praise, blame, and other reactive attitudes. De Waal himself concedes this point in a way, when in his conclusions (pp. 209–18) he argues that though some animals act co-operatively and can empathize with each other, they do not possess that form of morality which depends on what he calls 'cognitive empathy' (i.e. actually thinking about the feelings of others) and the explicit formulation and internalization of rules and possession of an abstract sense of justice, though they do on occasion engage in revenge, peacemaking, role-establishment, and succouring each other.

Davidson and Dennett, I apply mentalistic language to other people (and animals) in order to allow me to cope the better with certain types of organism. Regarding another person (or a dog) as having specific beliefs, desires, and emotions enables me to predict his or her future behaviour far more easily than if I had to rely on some as yet unknown combination of physics, biology, and neurophysiology. Presumably in attributing mental states to others I will also be able to attribute mental states to myself—and will do so.

The analogy between the intentional stance view and our recent speculations on consciousness is unsurprising because both accounts are attempts to show the functions of what are, after all, related phenomena. Mentalistic language is, indeed, simply the linguistic counterpart of consciousness. Self-consciousness, as involving explicit, determinate, and reflexive awareness of the contents of one's consciousness, requires some means of representing to oneself what those contents might be, and as we have seen in Chapter 3 language is vital for this purpose. There is no suggestion that the empathizing baboon or chimp has more than an intuitive sense of his own feelings or self, or those of his colleagues and rivals, or that he consciously attributes to them or to himself specific mental states. For this language is necessary as we have seen. But the functional similarity between the unreflective, unselfconscious empathy and the conscious intentional stance which subtends a concept of self on the part of the participant is striking. Both can not unreasonably be seen as facilitating one's dealings with others. Both unreflective consciousness of one's feelings and reflective self-consciousness give clear evolutionary advantages to their possessors, in enabling them to anticipate the actions of rivals and colleagues.

I have reservations about seeing the intentional stance as merely an alternative way to that of the physicalist of predicting the behaviour of organisms. Nor do socio-evolutionary speculations show *how* either consciousness or, even less, self-consciousness comes about. It may be possible to correlate consciousness in animals and self-consciousness in humans with certain levels of brain organization, and that might be as far as explanation can go, though for many this will serve only to pinpoint a mystery rather than to solve it. It is nevertheless suggestive to follow Humphrey in seeing a link between primitive awareness of one's own feelings and those of others and sociality. Doubtless, degrees of self-awareness have increased bit by bit, but the development of intentional language has been crucial to the possession of a robust sense of self as a possessor of beliefs and values which can in a

sense be detached from oneself through objective expressions, so as to
be examined and assessed by a scrutinizing self.

However, even before we reach the level of fully rational self-
consciousness of adult human persons, our sympathies are wont to
extend themselves to animals, just as animals may do to each other. We
do not just believe in an abstract way that it must be like something to
be a dog, a cow, a bat; when encountering these and similar creatures,
we readily and unreflectively feel with them and for them. One could
indeed follow the Humphrey line and speculate that projecting ourselves into the inner lives of animals gives us advantage in hunting
them, training them, and defending ourselves against them. None the
less, in training a dog or a horse, or defending ourselves against wild
beasts, we do not think of the animal we are encountering as a self,
capable of feeling and expressing self-involving emotions such as
pride, shame, anger (except as an unmediated and unthought response
to provocation) or love (except cupboard love, which is not love at all).
Nor do we have any sense in our dealings with animals, even with the
higher primates, that there is any yearning on their part that we
acknowledge or confirm their existence as selves.

The Gaze

It is, though, central to self-consciousness that the self looks for recognition as a self from other selves. As Sartre has argued brilliantly and
persuasively, in chapter 3 of *Being and Nothingness*,[15] central to the
whole dynamic here is the gaze; something not taken into account by
the functionalist accounts of the likes of Dennett. In speaking of the
gaze, Sartre refers to the way in which being observed in an activity by
another human being releases the possibility of a whole raft of emotions inappropriate were only an animal or inanimate objects present
(even including supposedly 'intelligent' objects, such as computers).
The gaze of another self-conscious being (who, like me, must be
embodied for the gaze to be directed on me as a physical, but more
than physical being and for me to respond to the source of the gaze)
pulls me up short, makes me conscious of my own self, and at the same
time conscious of the self who is not me. As Merleau-Ponty has
pointed out,[16] a dog's gaze would not affect me in the same way as the

[15] J.-P. Sartre, *Being and Nothingness*, trans. Hazel Barnes (Methuen, London, 1969).
[16] Maurice Merleau-Ponty, *Phenomenology of Perception* (Routledge and Kegan Paul,
London, 1962), 361.

gaze of even the most unappealing or undeserving stranger such as the 'hungry and homeless' vagrant in the tube. Sartre comments that in the gaze it is 'never eyes which look at us', it is the other—as subject, something which cannot be directly seen, but which none the less pierces us to the root of our own being and brings us to acute recognition of our own subjectivity precisely through our recognition that there is in the world a subjectivity other than our own. As Sartre puts it 'if this gross and ugly passer-by shuffling along toward me suddenly looks at me then there is nothing left of his ugliness, his obesity, and his shuffling. During the time that I feel myself looked at he is a pure mediating freedom between myself and me.'[17] Anticipation of punishment, remorse for an action which has provoked pain, and refraining from punishment-provoking behaviour: all these are witnessed in the animal kingdom. What we do not see is concern for the opinion of others, irrespective of any further effect: or any sense of the poignancy of simply being watched by another who has the linguistic resources to judge and who in watching brings me catastrophically to a sense of who I am and what am I doing. In being gazed at by another, my own self-consciousness is stirred acutely, but in such a way that I simultaneously realize the self-consciousness of the one who is gazing at me. I realize that, like myself he is an intending, reasoning, appraising, emotional being. My response—or his—can of course be the attempt to objectify myself or the other: to turn him into a thing, a living tool or a sexual object or to turn myself into a thing for him. In the sexual realm this involves oscillation between the poles of sadism or masochism, or in the human community more widely, between those of the master and the slave.

In his initial characterization of the gaze, Sartre describes a man unselfconsciously engaged in some slightly shameful act, such as spying through a keyhole, and then suddenly becoming aware that he is being observed. In this realization the whole world and his relation to his own activity is subtly but radically transformed. From regarding what is around him as simply a means to his end, and his own activity as no more than a way of fulfilling that end, once the spier is spied, shame reveals both the world and his activity as charged with an alien and unpleasant meaning. He sees the world not as simply there indifferent to judgement, at his disposal, but as a focus of evaluation and a source of affective attitudes. His own activity takes on a value in and for itself, as something shameful in this case. I and what I do are seen

[17] Sartre, *Being and Nothingness*, 276–7.

as another sees them, and I am jolted into a sharper level of self-consciousness than when I was unreflectively absorbed in my task. Actions and situations become meaningful in themselves, and not just stepping-stones to other needs or desires.

For Sartre, the level of self-consciousness revealed in the experience of shame depends on the existence of a community of beings who can provoke in each other a sense of the world as coloured by human affections, judgements, and norms. The experience of shame implies that others can judge me and that I can judge myself, and that the world I inhabit is not purely a world of objects, to be used or not as I wish, but that I am in a world suffused with human meaning. For Sartre himself, the meanings and implied judgements are founded on the gaze, something as he implies, mute, indeterminate, and unstable. His world is one of sinister and threatening observers creeping up on each other and compromising the innocence of some utilitarian action. We are like fish glassily and silently revolving around each other, and from time to time shaking each other into a form of self-realization which involves seeing ourselves as an actual or potential object of another's gaze, and so eating into our freedom of action and, in Sartre's view, of thought.

But the freedom of thought involved in solitary activity is one in which I am not fully and reflexively self-conscious. I am not yet aware of myself as an object, as an agent. I am a locus of feeling and of activity, and I have certain feelings and experiences connected with that, but I do not yet see myself as objectified, as part of a world in which what I do and think may be the subject of concern and evaluation on the part of others, and also to myself to the extent I can distance myself from the immediacy of my activity, taking up a perspective on it such as an outsider might have. For Sartre, the source of this type of decentring is the gaze: the look of another at me, combined with the fact that I am troubled by this.

The trouble that the gaze of another causes me suggests that it is not a look as such which is important. After all, being seen by a dog in a ridiculous or embarrassing situation is not enough to provoke in anyone feelings of indignity or shame. A look alone is not enough to do this, but only a look combined with judgement. There is an essential normativity in the gaze, which makes it hard to see how the phenomena so acutely described by Sartre could exist outside a community with the ability to express and discuss the relevant judgements. As Sartre himself says, with the look in isolation we have a 'pure indetermination in the game of possibles', and it is only when we are in direct connection with the other by language that we learn precisely what he thinks of us.[18]

Sartre is right to emphasize the affective character of our consciousness of ourselves or our concomitant consciousness of others. In being aware of myself or of another, in responding to his or her gaze, or in fixing our gaze on the other, we are not engaging in a purely intellectual or theoretical exercise. We are emotionally and motivationally implicated in various ways. Our feelings are engaged and exercised, we are impelled to act in response. In the pre-theoretical awareness of ourselves and of others provoked by the gaze, there is no disengaged scientific objectivity. In the world of human activity, a dispassionate look is a cold look and has its own emotional charge.

But emphasizing the emotional resonances of the human world does not mean that the human look could be the look it is outside a linguistic context. What the gaze does is to extract us from the world of things into which we have fallen and drag us into the world of interpersonal meaning and value—but *that* world is necessarily a linguistic world.

The Linguistic Community

It is a linguistic world because any world in which judgements are determinate and stable enough to be available for interpersonal negotiation will be a linguistic world. This conclusion has already been anticipated in Chapter 3, and is also familiar enough from the volumes of discussion and commentary surrounding Wittgenstein's private language argument. According to this, outside a linguistic community united by its sense of the meanings of its words, meaning and judgement would become irremediably unstable. Whatever I think or judge and whatever I take a word to mean is going to seem to me to be right; without others able to corroborate or correct what seems to me to be right, there is going to be no distinction between a judgement or use of a term seeming right and its being right. But how can the other even know what is seeming to me to be right unless I can give him some token of what I think, a token which at the same time fixes my meaning for myself? And this implies language or some analogous sign system.

A non-linguistic creature can, of course, respond to stimuli which we, as language users take to be, say, blue or painful, and so take the creature to be responding to blueness or to pain, but a behavioural response to stimulus is different from a judgement. For one thing, it need not involve any conscious awareness that the response is a response to a particular property or quality in one's environment. One responds as one is trained or programmed to do, but what leads up to the response need not enter one's consciousness at all. There is all the difference in the world between crying out in pain, which is unthinking and immediate and has no conscious lead up, and saying that one is in pain. In the latter case, one is bringing the feeling under a particular concept, a process which at some stage or at some time during learning or exercise involves conscious reflection as to the appropriateness of the concept to the type of sensation to which one applies it. Further to its being grounded in the conscious awareness of its maker a judgement can be adjudged true or false—indeed it is of its essence that it should be so judged—whereas a behavioural response simply happens or fails to happen. If it fails to happen when we expect that it will, then we speak of some malfunctioning of the organism or the mechanisms, which is what it is when someone utters a false judgement.

If these considerations carry weight for judgements such as that something is blue or that I am in pain, how much more will they for judgements relating to my character and behaviour? Not only do I here above all need help from the objectivity which can come only from distance. Self-appraisal will be at best incomplete, partial, and biased without at least the possibility of the other, but for the other to help with my self-awareness, he will have to be able to discuss it. Further, as Sartre reminds us, it is only the fantasy of the other which sparks us into that mode of experience, in which I see myself as the subject of interpersonal feelings; but, as we have already suggested, it has to be another who can do more than just look. And finally, the subtleties of self-appraisal—whether, say, one is feeling shame, as opposed to guilt or remorse—are hardly imaginable outside the linguistic context in which the various relevant distinctions can be drawn and discussed.

Full self-consciousness, then, presupposes that the self-conscious agent is a member of a community of other agents, who can judge about each other, and as a condition of their judging, express and convey their judgements linguistically. That we come to an understanding of what and who we are through the mediation of public concepts and in negotiation with others in our community may seem an intolerable restriction on our ability to define ourselves and to make our own way in the world, on our freedom, in other words. So, indeed, did it appear to Sartre himself. He resented the way our existence as free, self-defining beings was inevitably constricted and channelled by the judgements of others and by the way our identities were given to us by the social positions and roles which we inherit and subsequently, more or less reluctantly, accept. For Heidegger, too, the impersonal public world and its judgements are necessarily inauthentic, smothering the unique personality of each one of us in a pall of objectivity from which we only rarely ascend to a realization of true subjectivity and individuality through momentary and occasional fear of death.

Heidegger and Sartre are, of course, concerned to emphasize the subjective aspects of our existence in contrast to the objective, but in doing so they lose sight of the extent to which subjectivity, at least in so far as it is self-conscious, depends on a public world and its categories. For we are not in a position of Cartesian selves, fully conscious of our thoughts and internal states prior to our acquisition of a public language, and independently of corroboration of our use of language and judgement from other users of that language. The price of anti-Cartesianism regarding the mental is acceptance that my subjectivity is penetrated by the categories, thought-forms, and judgements of the public world. It is in the public world that I learn the language and categories which are applicable to the internal, subjective states of others and of myself, and I also learn that the truth of such judgements is intimately connected to public behaviour and performance. A man who is ashamed will behave in certain typical ways, and so will I if and when I too am ashamed. When I am caught spying at the keyhole, I may not blush (though I may); but my shame is likely to manifest itself in increased furtiveness, more care in the future, and perhaps even in refraining from such shameful acts in the future. Without accompaniments and a context of this sort, and the whole panoply of what Wittgenstein calls stage-setting, it is hard even for me to see my private feelings as feelings of *shame*, however intense they may be.

Mutual Recognition

What I have been suggesting over the last few pages is that one evolutionary advantage of self-consciousness for human beings may be that they are able to predict each other's behaviour better and so be more successful in the competition and co-operation of social life. But, once established, this type of mutual fellow feeling leads to demands for reciprocity unknown in the animal kingdom. Thus is introduced a need over and above those for survival and reproduction, and, as is suggested by Hegel in the dialectic of the master and the slave,[19] a need which will be unfulfilled so long as it is interpreted in terms of domination and submission.

A conscious being is a desiring being. From the biological point of view consciousness may be no more than an adjunct to the achievement of desire. But a self-conscious being is not just a vehicle for desire and a recipient of sensation; it is also aware that it desires, aware of its desires and feelings. Like consciousness, self-consciousness helps us achieve our desires. Hegel's crucial insight is that self-consciousness cannot rest content with fulfilment of desires. As Sartre shows, building on Hegel, the Hegelian insight is firmly rooted in everyday emotional and moral experience. Self-consciousness cannot rest content with the fulfilment of those desires which amount to purely physical satisfaction. Self-consciousness looks for recognition and validation by other self-consciousnesses. The personal in us cannot be satisfied by the impersonal, but requires recognition by other persons. In self-consciousness we move from the mechanical, even though flexible, operation of desire and its gratification to one's existence being regarded as a source of value. But the value of my actions and feelings cannot be a value for me only. For them to be a value for me, I must find my valuations endorsed by others—or there is no difference between fulfilment of desire and what is really good. The difference between Sartre and Hegel lies in their relative assessment of the permeation of subjective interiority by the categories and judgements of the public world.

As Hegel puts it, it is in the mutuality of self-consciousness that 'consciousness first finds its turning-point, where it leaves behind the colourful show of the sensuous here-and-now and the night-like void of the supersensible beyond, and steps out into the spiritual daylight of the present'.[20] The spiritual daylight of the present is more than the pure satisfaction of desire. In satisfying a desire for food, say, I gain the object I deserve, but in gaining it, I cancel out its otherness, it becomes part of me and no longer opposed to me. My awareness of self, which depends on a sense of distinction between myself and the world, thus subsides in an unthinking satisfaction.

---

[19] In G. W. F. Hegel, *The Phenomenology of Spirit*, trans. A. V. Miller (Oxford University Press, 1977), 111–19.

[20] Ibid. 110–11.

True or full self-consciousness can be maintained only in tension with another self-consciousness responding to me, interacting with me, even opposing me actively. But the spiritual daylight of the present of which Hegel speaks cannot be the relationship of pure domination and dependency represented by the master and the slave, nor can it be the purely exploitative system of reciprocal altruism, even though both master and slave and reciprocal altruism may prepare us for habits of behaviour which will eventually lead to a society basking in spiritual daylight.

The master is initially a person seeking recognition and endorsement of his value from another who is himself more than merely passively uncooperative, who is in his turn also self-conscious, and also seeking recognition from other self-conscious beings. The other equally desires recognition, and the two engage in a struggle of wills, to enforce recognition. He becomes the master who is prepared to risk more in the struggle, up to death itself. He thus shows that he is prepared to give up all, to be a pure self-consciousness or will, rather like the Sartrean ego who rejects the inauthentic judgements of others or of the public world. The one who is now master is untrammelled by any elements of his being inessential to recognition as a pure self-consciousness. Of course, both master and slave are faced with that fear of death which Heidegger sees as the source of individuality and authenticity, but as will emerge, fear of death is not enough to sustain any enduring sense of self. Through his assertion of will, though, the master achieves the recognition he craves, and gets his self-respect through the unconditional and total obeisance of the slave, an obeisance uncontaminated by empirical conceptualization. The master is obeyed purely as master, and not in respect of any other qualities he may have.

But the self-respect of the master proves to be unstable. The slave values the master only in virtue of the master's superior strength and will. And this is essentially negative, at least as regards any other qualities the master has. The *condottiere* or the robber baron may patronize the arts, but this is later and secondary, and is not the basis of his power. Moreover the master depends on the slave, whose respect for the master is not grounded in any qualities the master may have apart from his power. The recognition the master gains is one-sided and unequal, and does not rest on any genuine valuing by the slave of the master's activities. The master usually wants to enhance his self-esteem by doing something more permanent and admirable than winning a struggle. He may want a memorial, a piece of culture; but for this he needs the craft and skill of servants, employees, or even slaves.

The slave, by contrast, moves from the acute sense of self he acquired through his fear of death in the life-and-death struggle to something more positive. He begins to work, for the master, it is true, but in working he begins to order and discipline his desires and also to express himself objectively in the world through his work. He begins to move to a situation in which others can recognize him, not as pure desire or will as the master is recognized, but in and through his work. 'Work is desire held in check, fleetingness staved off', as Hegel puts it, a step on the road to an independence from desire and also towards the unforced recognition which the master never achieves. Or at least, the master is recognized through those he caused to work for him. The name Malatesta is remembered today, but precisely because of Sigismondo's employment of Alberti, Agostino di Duccio, and Piero della Francesca, not because of his successes against rivals in north-east Italy.

What self-consciousness looks for, then, is an unforced recognition from the other, one which respects me for the qualities which make me what I am, which is not forced from the other, which thus (far more than the unrespectful cringing of the slave) corroborates the value I would find in myself. Ideally, I also want to be in a situation in which I have the self-mastery that comes through discipline and which enables me to make a mark on the world around me, a mark which can be appreciated as such by myself and others, and in which I and they find me.

A number of important points emerge from all this. One is that a consequence of our existence as self-conscious persons is that we look for freely given recognition of our standing, qualities, and achievements from other self-conscious persons. A second is that the master not only largely fails in this, but he is subservient to the slave in not having the opportunity to express his freedom in the world by any creative activity. A third is that what we are and what we seek to have recognized as self-conscious selves is not primarily an inner self or set of transient inner states. Hume is quite right in failing to find anything constant within; further, when I want to be recognized and valued it is not as a bare Kantian transcendental unity of apperception, or even as the bare dominating will of the Hegelian master, let alone as the empty and insatiable *pour soi* of existentialist fantasy. It is rather as a person with some sort of stable identity given by projects and stable relationships in a public world in which the achievements and concrete individuality of each gain universal recognition. It is as a person, as a member of a civilized realm of work and mutual recognition, one perhaps founded initially on naked desire and naked struggle, but one transcending such impermanent manifestations in the more enduring forms and projects of civilized life. In Hegel's terms, one becomes an object of knowledge for oneself only in becoming an object of knowledge for others, a point grasped by Heidegger and Sartre, but grasped in purely negative terms. Compared with the Hegelian analysis, theirs is the revolt of the adolescent spirit against the world of objective order.

What we have here is an oscillation between an abstract need for recognition and self-expression and the concrete and particular ways in which this need can be fulfilled. We want recognition from the persons we encounter and at the limit, from all persons, but the forms in which we can be recognized most fully effectively, and also be most fully realized, are particular. I am a rational self-consciousness, even a Sartrean *pour soi*; but so is every other human being, and to be recognized in these terms hardly addresses anything specifically mine. As pure rational self-consciousness, I become at most a bearer of abstract rights, a mere place-holder in some theoretical legal economy. However inalienable might be the rights so accorded me, exactly the same rights must be accorded everyone else. In this system, I am precisely as much and as little as everyone else. But, humanly speaking, I am not simply a Kantian noumenal self. Of course I am that, if by noumenal self we mean rational agent, capable of free choice and responsible for my thoughts and deeds. As an agent of this sort, I will expect from others that recognition and respect which will allow me space to exercise my rationality. But my rationality, my noumenality so to speak, is itself something which comes to be more than a bare potentiality only in my actual existence in an actual human and linguistic community. Entering into social and linguistic relations with others is the way by which paradoxically enough, one escapes the determination of one's social context. Making the types of judgement characteristic of human language and self-consciousness implies that the one doing the judging is answerable to norms which are not constrained by any particular culture or belief-set. At the same time, in so far as my identity is constructed and situated historically and particularly, what I desire is for the respect due to me as a pure rational agent to be manifested as respect for the particular individual I am. I am a particular individual, with a particular heritage, commitments, and particular talents. I want these recognized as well as my abstract Kantian personhood. Even more fundamentally, I come to understand what I am in having my particularities recognized and reinforced by others.

Part of what civilization involves is the public recognition of my particular talents and projects. I move from individualistic self-assertion and (at best) the recognition of my physical dominance and courage in the face of death. I also move from the pure abstraction of universal human rights. I move towards a genuine unforced recognition of what I am and what I do, what might be thought of as rational endorsement from other rational beings, but endorsement of me as a particular individual, a possibility Sartre is unable to entertain owing to his entirely abstract characterization of individuality as pure transcendence of the given. If the dialectic of the master and the slave suggests that the first moment of civilization and of the supremacy of the rational is fear of death, it also suggests that it is through mutual recognition of our individuality in its full concrete particularity that civilization sublimates and transforms that fear. It suggests, too, that there can and should be a point at which, as Hegel argues in his *Philosophy of Right*, subjective self-seeking turns into a contribution to the needs of everyone else. That is, in a society where free exchange is possible, I satisfy my needs just in so far as what I do is recognized and valued by others; this is a 'dialectical advance', in Hegel's terms, in which in earning, producing, and enjoying on my own account, by that very fact I produce and earn, Hegel says, for the enjoyment of 'everyone else'. Even if this is an exaggeration, and what I do is not for the enjoyment of everyone else, at least as far as I am successful, it is recognized by a public impersonally considered, in which we are interdependent in an impersonal way. I may fail to get the recognition we want—of the noumenality-in-this-specific-individual—and I may fail to accord others the recognition they in their turn crave. But it remains a desire founded in our nature as rational *because* social and in our sociality. Its non-satisfaction through lack of unenforced recognition from other agents is a tragedy for those whose lives are unfavoured in this way, but a tragedy that has little to do with the non-satisfaction of the Darwinian drives to survive and to reproduce.

Self-Consciousness and the Ethical

The Darwinian world is one of struggle. The Darwinian world is the world of nature. Our current awareness of its interlocking and holistic nature notwithstanding, the Victorians were not wrong to see it as in many ways highly competitive and highly unpleasant for individuals, a life-and-death struggle indeed. Our current understanding of biology and evolution has, if anything, underlined the way that co-operation in the natural realm is ultimately self-interested in a narrow sense. So, too, is much co-operation in the human realm. Because of its evolutionary efficacy it is not implausible to think that we (and other creatures) have genetic predispositions towards sympathy and co-operation, and that these predispositions may form the instinctive basis of our ethical behaviour. But in the natural world, there is no imperative towards any form of mutual recognition beyond the unravelling of prisoner's dilemmas and reinforcement by strategies of tit-for-tat. Even if, as Michael Ruse has argued, our biology gives us a disposition to act on rules of co-operation, which we regard pre-reflectively as binding on us,[21] once we start to reflect, their hold on us will, from a rational point of view weaken, once we see them as no more than genetically based rules of behaviour.

What our Hegelian reflections on self-consciousness show is that, over and above any genetically based forms of behaviour, with self-consciousness comes a drive for mutual recognition of self-conscious beings through publicly assessible actions and norms. The dialectic of the master and slave suggests that at the level of self-conscious experience self-interest and mutual respect are not clearly separable. It also shows that so long as we remain at the level of domination and subordination and what might be called self-centred self-interest, the master is diminished as much as the slave. The suggestion is that because of the dynamic of self-consciousness, human fulfilment demands that sense of mutual respect characteristic of the ethical. The ethical does not demand sacrificing one's own interests so much as a realization that, given self-consciousness, self-interest goes hand in hand with a respect for the rights and achievements of others. The free-rider, the cheat, is in the end acting against himself, by subverting the very recognition he craves. In the natural world, the winner takes all, but in the spiritual daylight of the Hegelian forms of life winning is not all. Winning may indeed turn to ashes if achieved by ruse or subterfuge, and this is something everybody knows. With self-consciousness, the self becomes a public agent, seeking public approval and public shapes for its activity. In saying that a reason or a value which is not at least in principle publicly endorsable is irrational, one is not simply making a narrowly logical point about rationality, so much as a point about the type of world we enter in virtue of being self-conscious.

In saying that self-consciousness is essentially other-regarding, I am

---

[21] In e.g. his *Taking Darwin Seriously: A Naturalistic Approach to Philosophy* (Blackwell, Oxford, 1986), 222.

not saying that it marks a complete break with biology. After all, even in the animal kingdom, there is the phenomenon of care for the young, and in the case of human beings, it is plausible to suppose that a degree of empathy with our fellows is in-built, even with sets of epigenetic rules primitively governing behaviour. Indeed, in the speculations already alluded to, empathy and self-consciousness can both be seen as subserving the biologically useful function of providing agents with insight into the motivation, and hence of the future behaviour of their fellows. With fully self-conscious existence, such as our own, the price, if one may so put it, of this potential for insight is submission to mutually acceptable standards of evaluation so that my sense of my self can be endorsed and corroborated.

What is it that one wants in the way of cognition as a self-conscious agent? One certainly wants recognition as a bearer of some basic rights, such as the traditional rights of abstract liberalism. These basic rights are those to life, liberty, and to property, inasmuch as having one's own property is the condition of any effective liberty. Property is the barrier between self and the outside world which preserves for oneself the space and security for the development of one's own projects. Slavery and tyranny exist to the extent that individuals' liberty and projects are subject to arbitrary interference from the whims of others (including, of course, those of the bureaucratic state). In this context, the classic defence of private property is correct: that it, and it alone, effectively enables individuals to plan and pursue their own way through life. But property is not only a *means* to that end. In building his house or growing his garden or in building his library or his collection of antique pistols or even of old tram tickets (as the case may be), an individual is expressing his own self and is, concretely, making his mark on the world. Private property, then, is both the condition of and constitution of the self-expression which self-consciousness induces us to engage in.

It is important to realize that while self-consciousness requires recognition from our fellows, it is something whose dynamic has at its heart an individualistic aspect. What is wanted is a recognition of *me*, in the objective world and by my fellows, to be sure, but of me as one particular perspective on and talent in the world. This essential duality—that of an individual making his or her way in an objective and communal world throws some light on the debate current in political philosophy between liberal individualists and communitarians. The individualists are indeed wrong if they see society in purely individualistic terms, as a network which emerges from a contract formed from the decisions of fully formed and rational human individuals (regarded as abstract rights-bearers). To think that is to overlook the extent to which the very notion of contract and, indeed, of rational individuals depends on a pre-existing social context, a point to which we will shortly turn. On the other hand, communitarians, who would some- how disparage individuality and individual rights in the name of a social whole or collectivity, overlook the extent to which self- consciousness makes individuals of us all, subtending satisfactions and goals and assessments which cannot be wholly social. The cultivation of individual aims and goals is not a feature of particular social arrangements, merely, even though individualism is more stressed in some societies than in others. Individualism is, as Hegel saw so clearly, implicit in the initial dynamic of our social arrangements: that of par- ticular individuals seeking recognition from others.

The *Iliad* is sometimes seen as portraying a society without any sense of individual self or responsibility, but while it may be true that certain types of rootless or critical individualism are not encouraged in it (as the Thersites incident shows), the whole story depends on Achilles' sense that his *individual* deserts and prowess are being ignored by his leader. Furthermore the characters of individuals such as Agamemnon, Menelaus, Nestor, Odysseus, Ajax, Hector, Paris, Priam, Helen, and Andromache are so clearly delineated against what is often supposed to be a closed social background as to be fully alive for us today, more than two and a half millennia after their construc- tion, that one might be tempted to argue that the presence of an unquestioned ethical and social background is actually far more con- ducive to the development of individuality of character than its absence. The absence of such a background may tend rather to the production of the Woody Allen character, forever unsure of what to do—of his identity, as he would put it—and constitutionally incapable of that consistency of thought and action on which true individuality is built.

If I am to achieve recognition and to be valued by others, we must assume a shared background against which the relevant judgements can be made. The collector of old tram tickets would, one imagines, be generally regarded as a harmless eccentric by most people, though in a rather worse light if his eccentricity began to take over his life and undermine more important goals and relationships. On a more plaus- ible level, the activity of duelling, which was at one time very much tied up with mutual recognition, is now outlawed in most countries, a fact which simply emphasizes the way in which standards for mutual recognition and evaluation are both changeable and go beyond the fundamental rights enshrined in abstract liberalism.

What I have been calling the basic rights—to life, liberty, and property—can provide no guidance on the question of the relative merits or demerits of, say, tram-ticket collecting or duelling, even less of the standards which might be proposed in these fields. Nor, crucially, can they determine whether an institution such as duelling is to be regarded as infringing the rights to life and liberty. While those rights would rule out some practices, such as murder or slavery, they would be silent on many of the issues which confront people much of the time. Crucially they would not even define or delimit the intension (or even the extension) of some of the key rights. Is, for example, euthanasia murder? Or (as Nozick avers) taxation by governments for welfare projects slavery?

Substantive questions such as these cannot be settled without a conception of what is good, worth while, and right in human life, which is richer than that of abstract human rights. Equally, the mutual recognition and valuation which are sought by self-conscious agents require a background of judgement in which virtue consists in more than respecting each other's abstract humanity. What we have to consider is the relation between what I am calling a more substantive view of the self and critical human reason. On the analysis I am offering, both are integral to self-consciousness. The one supplies the possibility of the community of mutual recognition sought by self-conscious agents, and the other is the substantive complement of the abstract rational scrutiny of beliefs and practices which self-consciousness also brings in its wake. What we have here is something analogous to what we saw with regard to our systems of perception and belief, that critical and rational self-consciousness attacks specific even rather fundamental beliefs and types of perception, while simultaneously convincing us of the need for some such inherited framework.

In the case of our moral and evaluative dispositions, it is plausible to suppose that our 'thick' conceptions of the good life are largely inherited. It is plausible to think this because, as Aristotle shows convincingly enough, explicit judgement in these matters is grounded in feelings and dispositions. In the upbringing of our children we teach them factual knowledge, but also complexes of dispositions and reactions regarding *how* they should behave and react. Indeed, even in adulthood our moral judgements and aesthetic evaluations will continue to be based in such habits, which provide the ground in which our explicit judgements grow. As with factual beliefs, reasons come quickly to an end, and we are left with the bare assertion—that, for instance, there is a table before me, or that cruelty to children is abhorrent; assertions rooted in our way of life, to be sure, and connected to many other things, but assertions none the less and hardly further justifiable without question-begging. In the case of moral and other evaluations, the rock-bottom nature of some of our judgements is not a cognitive matter merely, for the judgements are, crucially, the expression of habits of action and reaction. The question, then, is where do the habits in question come from, and how are they to be considered by rational beings? To put this another way, how, without lapsing into irrationality, are we to treat (discuss, modify, accept) our basic moral feelings and habits?

As already observed, some of our basic dispositions and habits are based in our biology, and even some inklings of morality may be due to atavistic feelings of empathy, as Humphrey and others suggest. None the less, we need to be clear as to the extent to which moral behaviour does not conform to anything suggested by evolutionary models, particularly as developed by sociobiology.

Sociobiology

According to sociobiology, values are constrained and determined by facts about our evolution. Because of this, it becomes hard to see how a leading proponent of the position, Michael Ruse, can legitimately castigate the social Darwinists for their 'vile practices'[22]—at least, not if they arose from a dispassionate analysis of the natural world and its continuing and constraining impact on human nature. After all, as we have seen Dawkins putting it, the natural world is extremely nasty indeed, and from the human point of view, we should seek to transcend it as fast as possible. But if this is so, what should we do about those tendencies we undoubtedly have to favour kith and kin, do down our enemies, ignore the starving, and let the weakest go to the wall, at least so long as they do not actually force themselves on our attention? More to the point, if we follow the sociobiologist's line, what *can* we do to counteract these in-built biological tendencies, even if centuries of Christianity and socialism have convinced us that we *should*?

[22] Cf. Ruse, 'Evolution and Religion', unpublished paper, presented at the Royal Institute of Philosophy's Schools Conference 1995. I am grateful to Professor Ruse for permission to refer to the article.

It is worth recalling at this point a significant aspect of the Darwin–Wilberforce controversy. Bishop Wilberforce was, of course, worried about whether we were descended from monkeys. As Ruse correctly notes, 'it is the status of our own species which is [his] sticking point'. But part of Wilberforce's rationale for so sticking was his fear of what he took to be the ethical consequences of Darwinian naturalism. It was (as the Marxists say) no coincidence that he was a Wilberforce, the son of William Wilberforce, the emancipator of the slaves. Samuel Wilberforce was concerned that Darwinism might give comfort to racial and other supremacists. In a comparatively mild form, in *The Descent of Man*, it did, and in a less mild form its social ramifications have, as Ruse notes, contributed to the shame of the twentieth century.

As is perhaps not well-enough known, in *The Descent of Man*, Darwin had some very rude things to say about 'the barbarous races of men', about the low intellectual and moral capacities of 'savages' and about their 'hideous ornaments and equally hideous music'.[23] (Darwin had not had our experience of Birtwhistle and electronic pop, nor of John Galliano and Madonna; else he would surely not have been so rude about 'savages' in comparison to civilized peoples.) Nevertheless, to palliate the horror his contemporaries might have felt at learning they were descended from apes, he wrote that

For my own part I would as soon be descended from that heroic little monkey who braved his dreaded enemy in order to save the life of his keeper, or from that old baboon, who descending from the mountains, carried away in triumph his young comrade from a crowd of astonished dogs—as from a savage who delights to torture his enemies, offers up bloody sacrifices, practices infanticide without remorse, treats his wives like slaves, knows no decency, and is haunted by the grossest superstitions.[24]

In the pages immediately before this robust display of racist antispeciesism, Darwin had advocated more care in human breeding, rather as his follower Galton, who was later to elevate such thoughts into a complete system of eugenics. Those inferior in body and mind are to refrain from marriage, as are the poor. Just because the prudent avoid marriage, if we are not careful the inferior will supplant the superior. Further, if mankind is to advance still further, we must remain subject to 'severe struggle' and 'open competition'. The most able should not be 'prevented by laws and customs from succeeding best and rearing the largest number of offspring'. For the good of our-

---

[23] Darwin, *The Descent of Man*, i. 142.   [24] Ibid. ii. 440.

selves and our species we must in general do for ourselves what we do 'with scrupulous care' in the case of horses, cattle, and dogs, that is, we must learn and apply the principles of breeding and inheritance.

Politicians nowadays are universally reviled for expressing such thoughts except in Slovakia and places east. Even a few tentative caveats about the hazards inherent in children being born on welfare are likely to bring a heap of enlightened opprobrium on the hapless legislator's head. And one dreads to think what would happen to the career of an academic who ruminated, as Darwin did, on a time, not too far distant (he said) when 'an endless number of the lower races will have been eliminated by the higher civilized races throughout the world'.[25]

There is, I think, some muddle in Darwin's own mind at this point. The higher, civilized races—those doing the eliminating—are those with higher intellectual *and* moral qualities; that is, those less infanticidal, more decent, kinder to their wives, less bloody, more benevolent, more sympathetic even to the most debased. There is no reason to suppose that Darwin would have countenanced active genocide, but it is hard to see how the severe struggle he advocates as a means of raising the stock could flourish among a people possessed of universal sympathy and benevolence.

Social Darwinism and Herbert Spencer get a much worse press nowadays than Charles Darwin and natural or biological Darwinism. Yet, Spencer, like Darwin, also foresaw a time when a more elevated society would arise, motivated by altruism and benevolence. On the road to this millennium he did not (as far as I know) advocate genocide any more than Darwin did. What he advocated was what Ruse would presumably term the 'vile' doctrine that, in Spencer's terms, 'organized charity' is intolerable because it 'puts a stop to that natural process of elimination by which society continually purifies itself'.[26] Darwin, it seems, just wanted to stop the poor breeding, but would not have refused them charity. Leaving aside questions regarding the efficiency of indiscriminate charity and its benefits and disbenefits to its recipients, something Spencer may have been prescient in questioning, it is not at all clear whether much, except tone, separates the social views of Darwin and Spencer.

While in these areas, tone may be all-important, it is even less clear that Spencer's more robust attitude to social affairs and advocacy of unfettered competition is not closer to Darwinism than the neurotically

[25] Darwin, *Letters*, vol. i, ed. F. Darwin (London, 1887), 316.
[26] Quoted by Ruse, in *Taking Darwin Seriously*, 74.

mild-mannered Darwin was himself. Nor is it at all obvious that the greater altruism which both Spencer and Darwin hope for in the future is actually consistent with a situation in which, in Darwin's words at the end of *The Descent of Man*, advance requires 'severe struggle', failing which we will collectively and individually sink into what he calls indolence.[27] All this is simply reinforcing the message of *The Origin of Species* that the 'one general law leading to the advancement of all organic beings is to multiply, vary, let the strongest live and the weakest die'.[28]

It is true that Darwin speaks of 'other agencies' being more important than the struggle for existence in developing the 'highest part of man's nature'. But, at this point, Darwin (and possibly Spencer) became somewhat un-Darwinian.

In *The Descent of Man*, Darwin is anxious to show continuity between human moral sense and what goes on in the animal kingdom. To this end, he regales us with anecdote after anecdote concerning self-sacrificial behaviour in the animal world. He argues, persuasively enough, that social co-operation exists in the animal, as in the human world. He also says that, in general, modifications in structure and constitution 'which do not serve to adapt an organism to its habits of life, to the food which it consumes, or passively to the surrounding conditions cannot have been acquired [by natural selection]'.[29] This is correct, by Darwinian and neo-Darwinian standards. Natural selection is about *individual* survival and reproduction. For obvious reasons, it cannot explain the persistence of genuinely self-sacrificial tendencies (because, as already noted, the self-sacrificers will not survive and reproduce as much as the non-altruists, and so will die out, even as the rest of their community celebrates and profits from their existence).

Now, as we have also seen, it is true that reciprocity can be good for individuals, on a you-scratch-my-back and I'll-scratch-your-back basis, or, regarding sanctions, on a tit-for-tat basis. All this has been extensively documented by Dawkins and others. It is also true—as Darwin emphasized, rather more than some of his successors—that in humans and higher animals, behaviour is reinforced and underpinned by the possession of appropriate feelings and sentiments which may, once embedded, be applied beyond the case when it is my genes I am benefiting. So we should expect to find emotions akin to sympathy among social animals, in order to ground reciprocal behaviour. These

---

[27] *The Descent of Man*, ii. 439.   [28] *The Origin of Species*, 263.
[29] *The Descent of Man*, i. 90.

feelings may also be developed through the effects of sexual selection, whereby more sympathetic and 'caring', as well as better-looking, mates are preferred by females. But the magnification of sympathy in me cannot proceed to the extent whereby it prevents the twin goals of *my* survival and *my* reproduction, or those of my genes, if we are taking the modern sociobiological line here. For, if it does, I will die and fail to produce, and my over-sympathetic disposition will not be transmitted genetically to the next generation—so long, of course, as we are in a strictly Darwinian world.

What this means is that Darwinism, if consistent with the principle just enunciated about adapting *an* organism or its genes to its conditions, cannot invoke as an explanation for the persistence of a trait or disposition its *social* effects unless those effects also benefit the individuals involved, at least in the main and in the long term; for effects which are, as it were, merely social and have no tendency at least in the main and in the long term to benefit the individual will be de-selected by natural selection. Unfortunately and misleadingly, when it comes to discussing our intellectual and moral powers, Darwin introduces just the sort of societal account he should have eschewed. He talks about social instincts and the public good:

> It must not be forgotten that although a high standard of morality gives but a slight or no advantage to each individual man and his children over the other men of the same tribe, yet then an increase in the number of well-endowed men and an advancement in the standard of morality will certainly give an immense advantage to one tribe over another. A tribe including many members who, from possessing in a high degree the spirit of patriotism, fidelity, obedience, courage and sympathy, were always ready to aid one another and to sacrifice themselves for the common good, would be victorious over most other tribes; and this would be natural selection.[30]

This may or may not be true (Sparta? Genghis Khan? the British Empire? the Evil Empire?) and we are not told how long the relevant time-scale is; but what it is not is *natural* selection. The patriotic, faithful, courageous, obedient ones will—in natural selection anyway—be eventually outbred by clever drones and parasites who exploit their self-sacrificial qualities for their ends. Morality will decline and there will be no evolutionary advantage for individuals to buck the prevailing moral indolence and lassitude. One can see why Darwin is anxious to give us cases of dogs faithful unto death (even at the hands of their masters), of heroic little monkeys, and of patriarchal old baboons, but

[30] Ibid. i. 203.

this does not make the moral sense natural, if by that one means consistent with natural selection. In natural selection, such things will
have to be exceptions and will die out, unless preserved by artificial
means (such as in dog breeding). In natural selection, patriotism and
the rest will survive only if—at least in the main—they give the patriotic *individuals* some advantage. But when, as in World War I, patriotism and obedience were disadvantageous in a catastrophic way to
those in whom those virtues were most developed, the virtuous were
wiped out, and the virtues themselves became unfashionable and
impractical even until now. But there was nothing moral about this; if
patriotism, fidelity, and obedience were virtues then—and they surely
were—they are virtues still, even though natural selection may have
done its work and we, as a society are the poorer for it. But morality,
as opposed to natural selection, knows no such relativity to circumstance. Morality being general, universal, and categorical obeys the
Horatian edict: 'fiat justitia et ruant coeli' (let justice prevail, though
the heavens fall).

Michael Ruse understands all this, of course, which is why in his
paper on evolution and religion he raises his question about what he
calls the Love Commandment. What he questions there is whether the
Christian commandment to love thy neighbour as thyself is either
possible or desirable, given what he thinks we know about our origins
and hard-wiring.[31] Sociobiology builds on the Darwinian insight that
certain social and moral dispositions and feelings are hard-wired in us,
because of natural selection. While sympathy is one of these feelings,
it must, from a Darwinian perspective and in Ruse's words, be sympathy which involves a 'stronger sense of moral obligation to one's
immediate family than to friends, to friends than to strangers, and to
one's own country people than to foreigners'. We also doubtless have
other feelings hard-wired, which are hardly moral at all: anger, possessiveness, lust, and envy, to name but four. Will sociobiologists, such as
Ruse, also argue that all of this argues for 'a restricted understanding'
of morality? In so far as the feelings in question (*a*) lead to the survival
and reproduction of those in whom they exist, and (*b*) have a strong
evolutionary pedigree, it is hard to see how they could in consistency
fail to do so, even while perhaps, as Ruse does, speaking favourably of
sainthood and foreign aid.

Ruse seems well disposed towards foreign aid. But what he says
about morality makes it puzzling that he should. Darwin said this
about morality:

[31] 'Evolution and Religion' (see n. 22).

If, for instance, men were reared under precisely the same conditions as hive-bees, there can hardly be a doubt that our unmarried females would, like the worker-bees, think it a sacred duty to kill their brothers, and mothers would strive to kill their fertile daughters; and no one would think of interfering. Nevertheless, the bee, or any other social animal, would gain, in our supposed case, as it appears to me, some feeling of right or wrong, or a conscience. For each individual would have an inward sense of possessing certain stronger or more enduring instincts . . . an inward monitor would tell the animal that it would be better to (follow) one impulse rather than the other . . . the one would have right and the other wrong.[32]

And in 'Moral Philosophy as Applied Science' Ruse and his co-author E. O. Wilson find it:

easy to conceive of an alien intelligent species evolving rules its members consider highly moral but which are repugnant to human beings, such as cannibalism, incest, the love of darkness and decay, parricide and the eating of faeces. . . . Ethical premises are the peculiar products of genetic history, and they can be understood solely as mechanisms that are adaptive for the species which possess them . . . No abstract moral principles exist outside the particular nature of individual species.[33]

Conscience and right and wrong, then, are a matter of instinct, based on survival needs; no worrying here about duties owed to sentient beings, if we need them for our own survival. But equally, do we have an account of morality here at all?

Most commentators on the subject would reply to the effect that while the content of morality is doubtless constrained in a general sense by the physical and social circumstances of the beings in question, it is not so tightly determined as Darwin, Ruse, and Wilson suggest. That is to say morality, if it is to be morality, has to rest on principles of greater generality than what merely happens to conduce to my (or our) well-being at a given time. Going with Darwin's example, it is at least possible that *morally* bees ought not to kill their brothers and daughters (just as, in fact, Darwin himself objected to infanticide among the Japanese, despite their lack of *Lebensraum*). And perhaps we could say to Ruse that, whatever our kin-favouring hard-wiring, it is again conceivable that we ought morally to worry about the implications of the Love Commandment, and that morally it is no good excusing lack of concern by appeal to our *actual*

---

[32] *The Descent of Man*, i. 151–2.
[33] M. Ruse and E. O. Wilson, 'Moral Philosophy as Applied Science', *Philosophy*, 61 (1986), 173–92, at 186.

sentiments. If all this seems rather hypocritical, so be it; after all, hypocrisy is the honour vice accords virtue.

Actually, it is dubious that either Darwin (in the passage quoted) or Ruse (in his official position) have anything to say about morality at all, or, rather, what they have to say about it is tantamount to undermining it by removing its ground from under its feet. If morality is as context-dependent and group-interested as Darwin suggests, what sense does it make to think of morality as a system binding on all rational beings just in virtue of their rationality? Is our feeling bound by obligation no more than a trick forced on us by natural selection, in Darwin's case misleadingly and inconsistently to favour the well-being of the group?

Suspicions of this sort become even stronger when we turn to what Ruse has to say on the matter. Not only does he believe that 'there is no justification for morality in the ultimate sense'.[34] He believes that the prima-facie objectivity which attaches to moral notions is due to our genes prompting us in various ways. Rape, he says, is really wrong, and not just a matter of subjective sentiment. But what this means is that our genes get us in some unexplained way to regard the relevant sentiments and dispositions as objectively binding on us, because if we did not have what he refers to as the illusion of objectivity in their regard, they would fail to have power over us.

Both the example and the analysis are questionable. From a Darwinian point of view, we may indeed wonder what is so wrong with rape. Does not Darwin teach us that 'each organic being' is 'striving to increase at a geometrical ratio'?[35] Is the rutting stag worried about the consent of his mates or only about the strength of his rivals? Was behaviour among the Olympian gods or the tribal chiefs of legend so very different?

Ruse might be able to demonstrate that an instinct, or, more probably a social norm against rape can be justified in terms of reciprocal altruism together with the promotion of the selection of one's kin; and so with the other norms and practices which go to make up the content of our moral codes. And, further, that selection has favoured those individuals fortunate enough to live in groups operating under illusions of objectivity in this area. 'Human beings function better if they are deceived by their genes into thinking that there is a disinterested objective morality binding upon them, which all should obey.'[36] Where, though, is the *moral* force of this? What we are being pre-

---

[34] 'Evolution and Religion'.   [35] *The Origin of Species*, 129.
[36] Ruse and Wilson, 'Moral Philosophy', 179.

sented with is an account of the genesis of moral codes and sentiments, one which says nothing directly about their validity or bindingness on us from an objective point of view. Indirectly, of course, the account treats of their objectivity as illusory, while being sceptical about moral demands which cannot be seen as adaptive from a Darwinian perspective (i.e. as conducing to *individual* survival and reproduction).

Ruse draws one lesson from his story and it is one which will inevitably undermine morality. If morality is all he says it is, the illusion of objectivity will inevitably seep away once it is exposed as such, and morality will lose its power even as social cement. Human beings will cease to 'function better' when they realize that all this moral talk is no more than manipulation by their genes. Even less will people continue to respect the surplus of feeling epigenetically programmed into them. That is to say, sophisticated sociobiologists argue that the feelings induced in us to do good to our close kin are unable to make the precise calculations and discriminations regarding those we should sacrifice ourselves for, and to what degree—even though that is why they are in us and that, in the main, is their effect. They spread themselves over all those close to us, but by the law of averages, those most helped by them will tend to be those genetically closest to us. However, if we realized the truth of the sociobiological claims, wouldn't we then begin to be more discriminating, once we realized that there was no genetic reason for helping people who were not in fact our kin? More generally, the chances of a Darwinian account of what most people would see as morality are slim, as we have seen in the case of rape and the Love Commandment; even if such an account could be given, it would provide individuals with no compelling reason to be moral where they thought they could get away with immorality, whereas the very essence of morality is its unconditional nature and its non-relativity to circumstance.

But it is possible to take the very data Ruse examines and to come out with quite the opposite conclusion. Like Ruse, one could agree that people believe that moral demands are really binding on us. Like Ruse, we could agree that this belief in general and aspects of morality in particular are hard to square with a Darwinian standpoint on human life and activity. But far from using those observations to weaken morality's hold on us, could they not equally be used to suggest limitations on Darwinian accounts of human nature?

Ruse is insistent that the objectivity of morality is an illusion, presumably because nothing thrown up by our genes in the way he supposes morality is can be disinterested and objective. He also wants us to question adherence to the Love Commandment because it receives no warrant in sociobiology. But these conclusions follow only on the assumption that Darwinism is the whole story about our nature.

This claim itself needs careful examination. The theory of evolution does not pretend to explain every trait or disposition of an organism in a direct way. That is to say, it is not part of Darwinism to say that everything about an organism has selective advantage. Quite consistent with natural selection, changes in organisms can arise which have no particular advantage. An example sometimes given is that of the chin, which probably arose as a consequence of other features of jaw structure which did have selective advantage. This type of process, whereby advantageous features have non-advantageous features as by-products is known as pleiotropy, and is the basis of the distinction between selection of and selection for. So there was selection *of* chins somewhere in our evolutionary history, but not selection *for* them: the selection for was of the other features.

In addition to pleiotropy, other non-adaptive influences on selection include genetic drift, constraints due to the possibilities and limitations of bodily or organic form and the influence of chance. There is also, as we have seen, and as we shall see further in examining beauty, the important influence of sexual selection, in which for some reason a particular trait in a male is favoured by the females of the species. Slight initial favouring of this sort can set up a ratchet effect, whereby the descendants of the originally favoured and favouring have even more of the favoured property, and we end up with a prodigally luxuriant and not clearly adaptive phenomenon, such as the tail of the male peacock.

However, while Darwinism can and does encompass many elements which do not directly conduce to survival or reproduction, or do so only through the inexplicable influence of female caprice, what it cannot countenance are variations which are harmful. Every edition of *The Origin of Species* contains the following words, which may be taken as the fundamental axiom of the theory:

If variations which are useful to their possessors in the struggle for life do occur, can we doubt (remembering that many more individuals are born than can possibly survive), that individuals having any advantage, however slight, over others, would have the best chance of surviving and reproducing their kind? On the other hand, *we may feel sure that any variation in the least degree injurious would be rigidly destroyed.*[37]

David Stove, who quotes this passage in his article 'So You Think You

---

[37] *The Origin of Species*, 130–1.

are a Darwinian?'[38] points out that when we think of human behaviour 'any educated person' can easily think of a hundred characteristics seriously injurious to their possessors, concerning which, as he puts it, there is not the slightest evidence that they are in process of being destroyed. Stove starts off with the letter 'a': abortion, adoption, fondness for alcohol, altruism, anal intercourse, respect for ancestors, susceptibility to aneurysm, the love of animals, the importance attached to art, asceticism, whether sexual, dietary, or whatever.

Now, there is little doubt that some Darwinians would take all this as a challenge. The glove having been thrown down, they will produce epicyclical explanations as to why, for example, despite all appearance, anal intercourse, or respect for ancestors, or aneurysm do have adaptive potential. I can actually think of efforts in the first and second cases, but I will not expand on this here. Instead, I will mention some widely acclaimed virtues which have little to do with Darwinian success (survival and reproduction) and which are frequently injurious to their possessors in more than the least degree Darwin argued would lead to their being rigidly destroyed: feeding the poor, tending the sick, visiting the imprisoned, modesty, chastity, honesty, promise-keeping, integrity, respect for the rights of others, self-sacrifice, honour, and this is without even mentioning the theological virtues of faith, hope, and charity. Those virtues may presuppose another, non-Darwinian world for their intelligibility. It is worth remembering here that Kant, who had thought as deeply as most about such matters, found that even the ordinary common or garden virtues required another world for their ultimate justification. Kant saw all too clearly that in this world the wicked man flourishes under his bay tree.

The point is that neither morality nor virtue invariably reward those who practise them, and it cannot be so that they will. Contractarians and Darwinians can demonstrate that it is more pleasant for us all to live in a society governed by a moral code. Even or perhaps especially the wicked man, safe in the knowledge that those he deals with will respect his life and property, has a far easier time of it than he would have in a state of nature. But the temptation, to which the dedicated seeker after survival and reproduction will succumb, is to keep one's own promises only so long as one is part of a contract from which one is benefiting and which will be enforced. It is not at all clear that the reciprocal altruism of the so-called social insects amounts to more

[38] D. Stove, 'So You Think You are a Darwinian?', *Philosophy*, 69 (1994), 267–78. See, though, p. 179 below for a weakening of the claim from within contemporary neo-Darwinism—which is suggestive of the low empirical content of the theory.

than this; it is indeed hard to see how it could amount to more, given the Darwinian premiss that adaptive advantage is to be analysed in ultimately individualistic terms.

It is, indeed, not hard to imagine a society of sophisticated Darwinians, respecting various social mores, but only in the spirit and letter of a mutual protection pact, and heaven help those outside it. So we have altruism, but only to those we are in with or who can benefit us; promise-keeping, honesty, and integrity, but only to those we are or might be beholden to or who might pay us back, and similarly with honour and respecting the rights of others; the corporate works of mercy, but only in a restricted compass; modesty, but only from and towards our women, and chastity from my wife and other men, but from me not yet, oh Lord, not yet. Of course what we would have are not the virtues, but contractarian mutations of them, and misunderstandings all round, as we see in the arguments between Dawkins and his critics over the very use of such terms as 'selfishness' and 'altruism'.

These misunderstandings help the Darwinians to the extent that they hide from view the degree to which their proposals and analyses are revisionary of our normal moral consciousness. It is to Ruse's credit that he is open about this, but, as I said earlier, it is not yet clear what conclusions we should draw from his contention that Darwinism demands a restricted understanding of the Love Commandment, and, as I maintain, of much else besides. It might be argued that our practice owes more to Darwinian impulses than to morality, and that it is hypocritical to suggest otherwise. (Ruse does argue in this way.) But, even if this were true, it would not follow that the moralists were wrong to suggest that our practices ought to be less Darwinian and more altruistic, more committed to duty, honesty, and honour, come what may. And certainly, to go no further afield, in a society influenced by Hellenic philosophy, Roman moralism, and Christianity, such suggestions find ready acceptance among people across the whole political, moral, and religious spectrum. Might not this in itself show that Darwinism is limited as an analysis of human nature?

Thus, for example, mothers do not normally seek to have their children adopted by other women (as Dawkins says would be evolutionarily advantageous);[39] nor do sterile and homosexual humans normally devote their lives to the offspring of their brothers and sisters; nor, as W. D. Hamilton avers would happen in a society modelled

---

[39] Cf. R. Dawkins, *The Selfish Gene* (Paladin, London, 1978), 110.

on purely genetical considerations, do human beings sacrifice themselves for 'two brothers, or four half-brothers, or eight first cousins'.
Nor, I imagine, would anyone object if with a significant but otherwise ineffective change to all his genes, he and all his children were to
find themselves able to learn foreign languages more easily. What all
this goes to show is that in human life, we are not wholly or even primarily the servants or agents of our genes and their supposedly relentless drive to replicate themselves. Human life is far more plastic and
flexible than that, and, as we have urged throughout this section, subject to considerations not reducible to the evolutionary. Nor, with the
possible and much disputed exception of the Ik, is there or has there
been any human society in which *all* those with 'injurious variations'
have simply been killed off or allowed to be killed off, despite
Darwin's pronouncement that, in the struggle for life 'any variation in
the least degree injurious would be rigidly destroyed'.[40]

What we have to realize is that the Darwinian world is not just a
nasty world, as Dawkins, among others, is ready to concede. It is a
world which bears very little relation to any human world or society,
in which we do not find nothing other than genes and their bearers
striving and struggling for reproduction. That such urges and struggles
underlie some of our behaviour is not in doubt, and it is entirely possible that some of our feelings and habits may be based on reproductive strategies. The sympathy which underlies altruism may indeed
have had its origins in trying to put oneself in the other fellow's shoes,
as a strategy of attack. Reciprocal altruism may have set us on the path
to true altruism and kin-selection may have been the basis of feelings
of sympathy now far more widely applied. Self-consciousness may
give its possessor some evolutionary advantages, but it brings with it
social needs and obligations as well, as we have seen.

Human social life, then, is not Darwinian life. It involves behaviour
and habits which go beyond that. The question, then, arises as to the
origin and status of the traditions in which our non-Darwinian inheritance is embodied.

While admitting the difference between the human world and the
Darwinian one, it is certainly possible and not obviously implausible
to take a quasi-Darwinian approach to tradition, custom, and morality. That is to say, in face of the difficulties of relating moral principles
directly either to our biology or to self-evident canons of practical
rationality, we could look at them in terms of the survival and success

[40] *The Origin of Species*, 131.

of the societies in which they are embedded. In so doing, we will not
be suggesting, implausibly and incorrectly, that group selection can be
given a genetic foundation. We will, rather, be assuming for the sake of
argument that in the human world, individual behaviour can be constrained (non-genetically) by the traditions and conventions enforced
by groups. The proposal then would be that the rules or conventions
of particular groups can be justified and explained by the contribution
they make to the success of the group, which then, by sociological
rather than genetic means reproduces itself through passing on those
rules and conventions to succeeding generations. To anticipate what
we will say, in this case no more than in the genetic case can we justify
or explain individual behaviour in terms of purely group effects; but,
in order to give the evolution of tradition account a fair hearing, it is
important to realize that it is different from the group selectionism
rejected by contemporary neo-Darwinists in not suggesting a genetic
foundation for the selection of traditions.

A Social Evolutionary Account of Morality: The Evolution of Traditions

In *The Fatal Conceit*, F. A. von Hayek engaged in some speculations
which are highly relevant.[41] From the point of view of abstract rationality, Hayek argued, both traditional Christian morality and free
market capitalism seem defective. To rationalists and free-thinkers traditional morality, particularly on matters to do with sex and the family, has seemed and continues to seem otiose, burdensome, and
oppressive. Similarly the distributions of property and other goods
resulting from free market arrangements appear to many of those who
look at it in isolation to end up with results which are unfair and
irrational: the inequalities and waste and failures generated by capitalism are, to say the least, untidy. It must be possible to arrange things
better.

Against intellectualism of this sort, Hayek argues that those societies have flourished most which have been based on the family and
private property, that the apparently irrational taboos of religion have
protected these institutions; and that our ignorance of the workings of
social arrangements is such that what seems best to human reason
regarding morality and institutions is not likely to be best in practice.

---

[41] F. A. Hayek, *The Fatal Conceit* (Routledge, London, 1988).

To know what works best in practice requires the knowledge which comes only through experience, and experience has shown us no better way of organizing our affairs than via the family and private property. Or, to put it negatively, attempts to subvert or reform these institutions in the name of rationality and progress have had a far higher cost personally and socially than maintaining them, even without fully understanding them. So even if, individually and intellectually, some of us (or even all of us) jib at it, we should have the intellectual humility to bow before the lessons of experience and defend and preserve institutions which, whether we like it or not, are the basis of our success and civilization.

I do not want here to examine the truth of what Hayek says about the specific institutions of the family and private property. I want, rather, to examine the form of argument he deploys. What the people Hayek is castigating as rationalists are saying is that it is wrong to submit blindly to any tradition or institution. We must examine the traditions and principles we inherit and see whether what they are proposing seems reasonable, or not. Does the institution of property in the free market lead to unacceptable inequalities of wealth? Is conventional sexual morality oppressive to homosexuals or even to heterosexuals? If we decide that these are implications of traditional morality, then, as in any other field, we must use our intelligence and modify or abandon the principles. To do otherwise would show want of good faith, intelligence, or courage.

Hayek's response to the rationalist is to say that his reason, or even the reason of whole groups of individuals, is far more limited than is generally supposed. In large societies, no one can know the effects of policies or institutions in advance. The most we can know are their effects over long periods of time. As it happens the most efficient, most tolerable, and most civilized societies the world has known have been firmly based on the family and property, and if we are interested in the survival of Western democracy we disturb these institutions at our peril. In particular, the widely held intuition that capitalist economic arrangements lead to injustice is a throwback to economically more primitive societies, in which all individuals knew each other, and transactions were face to face. Well-meaning attempts to enforce in large societies intuitions of justice which are appropriate, if at all, only to small societies will have the effect of reducing the large societies in question to Third World conditions, as we saw in Eastern Europe under communism.

It could be argued that what looks like irrationalism in Hayek—

unquestioning acceptance of traditional institutions—is only a prima-facie irrationalism, an irrationalism at one level only. At a higher level of analysis—in the evaluation of societies as a whole—he is as ratio-nalistic as his opponents. What he is actually providing (whether the details succeed or not) is an indirect, evolutionary justification of insti-tutions and practices which we are (in his view) unable to assess directly. It is an evolutionary justification because he is saying that the institutions and practices have evolved, as part of the societies in which they are embedded. Like the societies themselves, their evolution has been unplanned and its effects unforeseeable. Nevertheless, they have survived in societies which have survived, and appear to be important in the survival of the societies in question.

In Hayek's own words

the (moral) tradition is the product of a process of selection from among irra-tional or rather 'unjustified' beliefs which, without anyone's knowing or intending it, assisted the proliferation of those who followed them . . . The process of selection that shaped customs and morality could take account of more factual circumstances than individuals could perceive, and in con-sequence tradition is in some respects superior to or 'wiser' than, human reason. This decisive insight is one that only a *very* critical rationalist could recognize.[42]

How are we to assess the final sentence? It could be intended ironi-cally, but it need not be so read. It could be taken to imply that while only an indirect approach to moral customs and values is possible, an indirect approach is none the less possible. Is that not, indeed, what Hayek is doing himself when he criticizes central planning on the grounds that it does not work? For despite his ignorance thesis regard-ing the effects of particular policies and institutions, Hayek does think we have grounds for assessing the comparative success of societies as a whole. These are, of course, in a broad sense evolutionary grounds, suggesting that in some circumstances societies of one type have out-performed others. The argument thus becomes a two-step rationalism. The first step is to observe that societies with free market arrange-ments and traditional family set-ups have triumphed over closed soci-eties of various sorts, feudal, aristocratic, and now communist or socialist. Therefore (second step) we should impose or promote in the population those values which are integral to the free market. (The nuclear family is important in all this because it provides both a focus of and a motivation for the widespread acquisition of property.)

[42] Hayek, *The Fatal Conceit*, 75.

It is precisely at this point, though, that an evolutionary approach to ethics becomes problematic. We are saying that certain societies with certain practices have succeeded. Can we, though, be sure that the success is because of the practices rather than despite them? Or even that in our identification of the practices we have picked out the elements which are important? Given Hayek's belief that the reasons why things succeed in large societies may be very different from the reasons people think they succeed, as Hayekians we surely ought to be less confident that we have answered these questions correctly and can distinguish just what in a society contributes to its success. It is worth remembering in this context that in biology we cannot infer from overall evolutionary success of a species to the efficacy of every faculty or organ of that species. It is, therefore, problematic in inferring more in sociology than is permitted in the safer world of biology. But even if we could be more confident than an evolutionary argument should allow us to be about the precise reasons for the comparative success of some types of society over others, it is not at all clear what conclusions this allows us to draw for the future. Here, as in biology, we have to remember that arguments from past (or even present) evolutionary success to future performance are extremely risky. The dinosaurs died out, and so, for all we know, might liberal democracy.

In the social sphere the risks are compounded by multiple uncertainties and the complexities of conditions, which make it hard to see situations being repeated precisely enough to allow for conclusions to be drawn safely. In the first place, it is far from clear what is to count as success in society. Hayek, not unreasonably, tends to concentrate on its ability to support ever larger numbers of people in reasonable comfort. One is reminded here of Adam Smith's adage about the greater wealth of the ordinary individual labourer and his family in commercial society compared to that of the African potentate surrounded by a thousand naked slaves. But this is an extreme example, between which there are many gradations, so many as to render this type of broad-brush comparison unhelpful in practice. We can all agree that an industrialized society combining some degree of division of labour and a market economy of some sort is far more productive and spreads wealth about far more than a primitive despotism. Outside universities and semi-religious groupuscules of the left, pretty well everyone agrees that centralized socialism is a failure, economically and morally. We might, then, say that social evolution has killed off two types of social arrangement, though even saying that points up another disanalogy between the social and the natural. Can anyone be sure that

versions of primitive despotism and state socialism are not still with us
even after the fall of the Berlin Wall, and that, un-dodolike, others will
not in the future emerge?

If evolution works in the social sphere it does so less ruthlessly than
in the natural. Or, to put it another way, nature learns the lessons of
failure better than human beings. Our freedom and rationality allow
us to repeat old mistakes where nature is less flexible and less tolerant;
or, rather, the freedom of some allows them to inflict old mistakes on
others who are either too weak to prevent it or too muddled to see
what is happening.

Let us, though, concede to the evolutionist that certain arrangements have survived better than others. It still does not follow that we
should uncritically replicate everything in the societies adopting those
models. Even on an evolutionary model, doing so might well lead us
to disaster by preventing the emergence of the new and modified
forms necessary to deal with new circumstances. To take a topical
example, in the West capitalism has tended to go along with the development of mass democracy. In the view of some apologists for capitalism, indeed, the two go together as the expression of individual
choice and power in the economic and political spheres respectively.
But, following de Tocqueville, there is a powerful argument to the
effect that mass democracy tends to lead to a number of disabling
social traits.[43] As individuals in mass democracy become materially
more and more comfortable through the economic prosperity provided by capitalism and the redistributive arrangements they demo-
cratically vote in, at the same time they expect the state to supply more
and more of their needs. The state thus becomes benignly more and
more powerful, and the individuals in it less and less enterprising and
less open to the risks that make further progress possible. If there is
any truth in this, then a society combining capitalism with mass
democracy may be in a state of internal tension, or even contradiction
(as exemplified by the current cant phrase 'the social market'); it may
be moving towards a decadence from which its infantilized and hedonistic populace is unable to resist the challenges of more vigorous and
less comfort-orientated competitors. And here it is surely significant
that the current stars in the economic realm are the market-driven but
highly authoritarian countries of the Far East. Might Malaysia and
Korea even now be providing us with examples of evolutionary devel-

---

[43] Cf. A. de Tocqueville, *Democracy in America*, trans. G. Lawrence (Fontana, London, 1994),
691–5 (the chapter entitled 'What Sort of Despotism Democratic Nations Have to Fear').

opment to challenge and eventually overwhelm our outdated models of progress?

Once again, it is important to underline that I am not saying that Malaysia or Korea are definitely models for the future. Even less am I saying that we in the West should imitate them. I am simply attempting to illustrate some of the difficulties involved in looking at social development and public policy from an evolutionary perspective. Even if Hayek is right in claiming that it is only the adoption of rules of property which allow some to get far richer than others which has permitted the material and social advances of the West, it clearly does not follow that these are the rules others should adopt in other circumstances. Nor does it follow that we should cling on to the rules unmodified in changed circumstances, or that doing so will guarantee our future success. There are, in any case, as already suggested and as one might expect of an evolutionary development, tendencies in liberal democracy pulling in opposite directions, and which may in the end prove its undoing.

In fact, Hayek himself does not suggest we should simply acquiesce in what has been. What he actually says is that 'the whole structure of (economic) activities tends to adapt . . . to conditions foreseen by and known to no individual'.[44] In one sense, this is perfectly true. For all sorts of reasons, including the unpredictable influence of future scientific and technological knowledge, we cannot predict the future course of history, even in outline. Nor can we know how the unintended consequences of the decisions and policies we embark on will work out. Future conditions, including those conditions determined by our own actions, are not known to us.

But from this important reminder of our ignorance and fallibility, Hayek seeks to derive a conclusion which does not follow. He says that *because* the information individuals and organizations use to adapt to the unknown is necessarily partial, and works through society through chains of signals individuals cannot follow, we should eschew what he calls 'deliberately planned substitutes' for self-ordering processes of adaptation. In other words, we must avoid conscious planning, particularly from governments and other central agencies, in favour of the spontaneous order which will result from working through the whole of society of market decisions taken by individual producers and consumers with only self-interested motives in mind. The thought here is that of Adam Smith: that the butcher who

---

[44] Hayek, *The Fatal Conceit*, 76.

sells me meat at the cheapest price (so as to outflank his competitors) does so with no thought of benefiting me. But he does benefit me, just as I benefit him by buying his meat. And a society whose economy is controlled by myriads of such decisions will tend to drag labour and capital to their most productive and desired uses, to overall benefit.

It is, I suppose, true that agents in biological evolution do not act for reasons to do with the balance or well-being of the whole, and that an order of a sort emerges nevertheless. But then—and particularly if the gene is regarded as the agent or subject of biological evolution—it is not clear that acting for reasons plays any non-epiphenomenal role in evolution at all. (As Dawkins says, in evolution we find 'the total prostitution of all animal life, including man and all his airs and graces, to the blind purposiveness of these minute virus-like substances' (genes).) It is hard to see what light a realm of blind purposiveness can throw on a world such as ours, largely resulting from creatures who can act with reason and foresight.

Obviously the individual entrepreneur and the individual consumer often act with reason and foresight, even if neither can know all the ramifications of his actions, and even if as a rule neither acts with the global effects of his actions in mind. Nevertheless each might sometimes think about how his actions affect 'the whole structure of activities'. An entrepreneur might engage in or refrain from engaging in a price-cutting war on the basis of what he believes will be its effects on his market as a whole. Individual consumers may make ethical or environmental decisions in their purchasing, and can be remarkably successful in affecting whole markets in so doing. Of course, in cases such as these, as in any other of their decisions, agents are acting with limited knowledge, and may be wrong. But the fact that one does not have perfect knowledge does not make it more rational to act as if one has no knowledge at all. Arguments of this form can sometimes appear persuasive to philosophers, as, for example, when it is argued that as one can never know which of two dying patients each competing for the one available organ will actually live longer or better, one ought simply to toss a coin to decide between the two. But this type of sophistry has been generally and rightly regarded as unhelpful by the practitioners who actually have to make the decisions, and do so with a good deal of care and personal difficulty.

Hayek might argue that the ignorance of conditions and effects is greater when it comes to central government and its policies than it is for individual entrepreneurs and consumers regarding their actions, but even if this were true, it would not show that acting on limited

knowledge were no better than acting randomly or not at all. And acting not at all is not an option for government in modern society for at
least two reasons. First, it has great centralized power already, so even
to divest itself of it, it will have to act from the centre. This, indeed, was
Mrs Thatcher's project in the economic realm and the project has, like
most policies, met with some success in its own terms, as well as some
undesired effects. Certainly, contrary to the expressed intention, it has
not disengaged the state completely from economic and social policy.
Secondly, it could be argued with plausibility that in modern societies
a certain amount of government regulation is always going to be necessary even to keep the conditions of a genuinely free market in
place—against anti-competitive cartels, multinationals, and trade
unions, who, often in concert will act so as to dominate and distort
markets, impeding what Hayek would regard as the free flow of economic information to and from consumers and producers.

So the argument from ignorance seems to have no more force when
applied to central government than when applied to individuals acting
on their own behalf, and, further, some involvement in economic policy would seem inevitable, even for a government committed to
Hayekian principles. We have to conclude then, that Hayek's insistence that societies composed of rational agents act as if they were
composed of unthinking genes, even if practicable, amounts to little
more than dogma, and unworkable dogma at that.

On this point, Michael Oakeshott writes that

a plan to resist all planning may be better than its opposite, but it belongs to
the same style of politics. And only in a society already deeply infected with
Rationalism will the conversion of the traditional resources of resistance to
the tyranny of Rationalism into a self-conscious ideology be considered a
strengthening of those resources.[45]

In other words, Hayek's opposition to rationalism has itself become
an ideology, an erection of principle where the genuine opponent of
rationalism would advocate a far more pragmatic, less inflexible
approach. To use central planning where appropriate and when it can
be monitored is more reasonable than a quixotic and unavailing
demand to avoid it altogether, just as making medical decisions on best
information is better than making them pretending one had no information at all.

Hayek's combination of an appeal to evolution with the attempt to
restrict the operation of human reason in politics and economics is, in

[45] M. Oakeshott, *Rationalism in Politics* (Methuen, London, 1962), 21.

any case, in some tension with the evolutionary doctrine of the irreversibility of evolutionary processes. That is to say, Hayek and others may regret the incursion of reason into our affairs; they also have valid points to make about its limitations, but they cannot have us pretend that what has happened has not happened.

Reason and Tradition

As self-conscious agents, there are occasions when we cannot avoid reflecting on our moral and political intuitions, and asking ourselves whether they are correct, just as at times we cannot avoid considering the truth of our factual beliefs. Particularly in situations of conflict of principle or belief, reflective discussion is one of the resources afforded us by self-consciousness. And if anything in our Hegelian reflections on the impulse of self-consciousness to mutual corroboration is correct, reflection and discussion with one's opponents will in the end be a more satisfying way of resolving disputes than conflict and the suppression of opposing views. What we are looking for is a position which is fair in the sense that it can be seen in ethical discussion as reasonable by *all*, an intuition long pre-dating Hegel and by no means confined to the idealist tradition. For Aristotle seeing ourselves as others see us is the beginning of ethical wisdom, while Smith and Hume take the fantasy of the impartial observers to be central to moral apprehension. In such contexts, appeals to the past success of practices will at most be part of any discussion, and cannot be used to close reflection down, as Hayek appears to want.

We cannot simply accept our moral and social values as the unintended consequences of an evolution we do not understand, as Hayek advocates. Apart from anything else, seeing morals and traditions in terms of blind traditions blindly accepted leaves it quite unclear why people did or do accept them. Hayek himself sees this as a problem. In the final chapter of *The Fatal Conceit*, he writes

How could traditions which people do not like or understand, whose effects they usually do not appreciate and can neither see nor foresee, and which they are still ardently combating, continue to be passed on from generation to generation?[46]

He says that part of the answer is through the evolution of moral orders through group selection—groups which behave in the appro-

[46] Hayek, *The Fatal Conceit*, 135–6.

priate ways simply survive and reproduce. But we are not dealing here with unconscious deterministic genes. We are dealing with human beings, who need reasons for their actions, and who can and do abandon practices if they do not like them, or see no reason for them. Social pressures can, of course, supplement individual motivation and punish individual backsliding, but even social pressures have to be enforced, and enforced at some level by people who believe in them. Even if the guardians see their noble lies as lies, the auxiliaries at least will have to believe them. As we saw in the case of the Soviet Union, even the most oppressive totalitarian structure will in the end collapse under pressure from without, if no one within believes in its myth. The point is that in the matter of belief and practice, group selection of beliefs and practices is insufficient to give individuals reasons, and beliefs and practices are believed and performed by individuals, who need in the end to assent to them.

Hayek, indeed, recognizes that his group selection explanation is insufficient, for he says that the unpopular and anti-instinctual practices responsible for liberal democracy were actually upheld by religion. This, in a way, only puts the question one stage further back, for the religious beliefs themselves would have had to have seemed generally plausible to get embedded in a society. *Pace* Frazer's claim in *The Golden Bough*, and for the reasons just adduced concerning morality, a 'conspiracy of priests' can only work on individual minds which are receptive to the dogmas in question (and will, in any case, fail in its work within a comparatively short period if the conspirators themselves are merely hypocritical). Precisely because we are self-conscious and reflective agents, group selection can never be a sufficient condition for the presence in a society or culture of particular sets of beliefs or practices. *Individuals* in that society have to be convinced or forced to adopt the beliefs or practices.

Group selection cannot explain individual acceptance over a society as a whole, which must come from individual minds consenting: nor can it explain the working of force in this area, for those doing the forcing will have to be convinced of the value of the beliefs and practices they are enforcing, which is something quite independent of any Darwinian advantage they might have. The enforcers will also have to convince enough of those they are forcing if the beliefs are not ultimately to erode and break down (a phenomenon frequently seen in history, most recently in Eastern Europe).

We might also question whether the mechanisms Hayek mentions for the spread of successful (or other) traditions are straightforwardly

Darwinian. He refers to migration and emulation, but if either is to succeed in spreading traditions of belief or practice, they will have to work through conviction of the emulators or the host population. And conviction here must involve belief in the truth and validity of the traditions, not just in their efficacy, even if manifest efficacy can, on occasion, be a motivation to accept their truth. To give a concrete example: as just mentioned there is currently much talk in the West of the economic superiority of the authoritarian collectivism and organicism of the Far East. But as far as our own practice is concerned, this is going to remain talk unless we can find for ourselves a totem in which we believe, around which to collect and play down our individual personalities and differences.

The same point as we are making in criticism of group selection of ideas can be made about Dawkins's postulation of memes.[47] These are supposed to be self-replicating ideas which virus-like seek to propagate themselves by colonizing and taking over human minds. Leaving aside questions as to the physical or genetic basis of memes, the key point to notice is that to the minds so colonized, the memes must seem rationally persuasive. In so seeming, the focus of explanation will move from the supposedly non-rational and purely evolutionary power of the meme to replicate itself to the extent and nature of the rationality of the individuals who accept the memes and who have a degree of critical freedom in regard to them.

Just as in Dawkins's scheme of things, biological organisms are prisoners of their genes, so our minds become prisoners of the memes which colonize them. But, whatever might be said about organisms and genes, minds precisely are not prisoners of memes. We might, as in advertisng, be irritatingly influenced by jingles and slogans; but our irritation is the greater precisely because we know that this is not the normal case, and that generally speaking we have the power (and even the duty) to appraise what we think and are influenced by on rational grounds. We can reject or modify ideas we find rationally wanting, and many of us do. And if we accept an idea after examining it and finding it plausible, our case is precisely not that of our mind having been colonized by a phrase one 'cannot get out of one's head'. One has begun to take a rational attitude to the idea, which presupposes a degree of reflectiveness to and distance from it, something which would not be the case with a meme which simply colonized one's mind without one having had any say (or thought) in the matter.

[47] Cf. Dawkins, *The Selfish Gene*, ch. 11.

Dennett, who accepts and develops Dawkins's notion of the meme firmly rejects this notion of reflective and rational control of our memes: 'the "independent" mind struggling to protect itself from alien and dangerous memes is a myth', he says, much as he had argued earlier that our ordinary concept of consciousness was a myth.[48] According to Dennett, the mind is 'to a very great degree' the creation of memes, 'an artefact created when memes restructure a human brain in order to make it a better habitat for memes.'[49] Dennett goes on to argue that the self is nothing over and above a complex system of interaction between my body and the memes which infest it and that the norms which we use in criticizing and developing some of our memes are themselves simply more powerful memes.

While there is some truth in what he says about the mind being initially set up in a social context and being filled with ideas and values it inherits and also in his claim that culture (memes) can outstrip biology (genes), Dennett does not adequately appreciate the drive to rationality and reflection inherent in self-consciousness. There are no limits in principle to the criticisms and improvements we can make to our ideas, and in making criticisms and improvements we correctly conceive ourselves on occasion to be operating under criteria which *are* rational and adopted *because* they are rational: criteria relating to such things as consistency and rules of evidence. Memes they may be, but they are memes of a particular sort and accepted by us precisely because of their key role in assessing the rational acceptability of other memes, which if they survive the test may themselves be regarded as rational or reasonable. Mistakes we may make, but as self-conscious we are driven logically in the direction of rationality. Appreciating this logico-cum-psychological drive shows how inadequate a picture of our mental life is given by the thesis that our minds are nothing but a habitat for colonizing memes, which then control us for reasons which have nothing to do with their actually being reasonable (as opposed to our being memetically conned into dubbing them reasonable). In the final analysis, if Dennett's analysis is true, we have no reason in reason to accept it and we should reject it. This is because as with every other belief or theory we should be minded to accept it only if we have

[48] D. Dennett, *Darwin's Dangerous Idea* (Allen Lane, London, 1995), 365, (the earlier analysis was *Consciousness Explained* (Little, Brown & Co., Boston, 1991)); with Roger Fellows I argued (in 'Consciousness Avoided', *Inquiry*, 36 (1992), 73–91) that *explaining* consciousness was just what Dennett failed to do because he attempted systematically to deny the very existence of the features of consciousness which make it both interesting and problematic, in particular the existence of conscious selfhood and its central importance for rationality and judgement.

[49] Dennett, *Darwin's Dangerous Idea*, 365.

genuine reasons to think it true—which is just what the meme analysis leaves no room for. So we should reject it.

The Traditional

While we are never simply passive in the face of memes or the beliefs or values of our group, what can exist or fail to exist in a culture is a mentality more or less favourable to the transmission and acceptance of traditional beliefs. Part of what Hayek is objecting to is the attitude criticized by Popper in his essay 'Towards a Rational Theory of Tradition' and characterized by him as follows: 'I am not interested in tradition. I want to judge everything on its own merits . . . quite independently of any tradition. I want to judge it with my own brain, and not with the brains of other people who lived long ago.'[50] A paradigm case of the man Popper is describing is Keynes, who, in Hayek's words believed that 'by taking account of foreseeable effects, he could build a better world than by submitting to traditional abstract rules'.[51] A self-professed immoralist, Keynes had nothing but scorn for what he disparagingly referred to as 'conventional wisdom'. Keynes's point was that it should be possible to come to a snap judgement regarding the worth of any purported value or principle by considering the consequences of adopting it. We could, as it were, forget the penumbra of tradition which might be surrounding it and consider it in a dispassionate rationalistic way, simply in terms of its effects.

Such an attitude will certainly be destructive of tradition, as George Orwell observed in 'Inside the Whale': 'Patriotism, religion, the Empire, the family, the sanctity of marriage, the Old School Tie, birth, breeding, honour, discipline—anyone of ordinary education could turn the whole lot of them inside out in three minutes.'[52] Orwell is not, of course, saying that anyone 'of ordinary education' would be right so to do. The point is that a rationalist, consequentialist approach to traditional values is likely to fail to see what is good in them, and to see only their immediate and immediately ridiculous or oppressive qualities. Such analyses are likely to overlook and misunderstand the long-term advantages of traditional values, in part because as Hayek

---

[50] K. R. Popper, *Conjectures and Refutations* (Routledge, London, 1963), 120–1.
[51] Hayek, *The Fatal Conceit*, 57; Hayek discusses Keynes's 'My Early Beliefs' in the latter's *Collected Works*, vol. x (Macmillan, London, 1972).
[52] George Orwell, *The Penguin Essays of George Orwell* (Penguin, Harmondsworth, 1984), 127.

rightly insists such things are very hard to discern before the values are challenged. We are only now beginning to understand the effects of extending Keynesian personal morality through the whole of society in the 1960s, and to appreciate the sense of the objections traditional moralists made in often blundering fashion at the time.

Another thing Keynes said of his early attitudes was this: 'I can see us as water-spiders, gracefully skimming, as light and reasonable as air, the surface of the stream without any contact at all with the eddies and currents underneath.'[53] Knowingly or unknowingly, Keynes touches here on an important aspect of rationalism. Destructive as it can be of traditional allegiances, it fails in a positive way to make contact with the things which actually move people (which is partly why, having severed contact with traditional allegiance, rationalist morality tends to be little more than an attempt simultaneously to satisfy the maximum possible number of personal, usually hedonistic, preferences). Once again, George Orwell's analysis is telling. Getting rid of such primal sources of value as patriotism and religion did not get rid of the need for something to believe in, even for intellectuals. In the 1930s for intellectuals that something was too often Stalinist communism, a swallowing of totalitarianism and talk of 'necessary murder' by those who had had direct experience of nothing but liberalism.

Morality and politics both in part involve negotiation towards ends which everyone would accept. Rationalist approaches to morality too often move too far from what actually motivates real people, who are, we should remember, products of evolution and tradition—on this, at least, Ruse and his fellow sociobiologists are correct. In an attempt to engineer impartiality it is all too easy to produce bloodlessness. In an effort to eradicate what seems to them to be irrational prejudice, rationalists produce systems which destroy local and historic allegiance, without producing any countervailing commitment or motivation. What seems to them should be acceptable to everyone remains a concept merely, valid on paper only. In practice, it is acceptable to very few. The irony of the rationalist position is that what is intended as a project to satisfy everyone ends up dissatisfying the vast majority of real human beings (who might actually find an *enemy*, someone prepared to fight for his or her national identity, more intelligible and worthy of respect as a human being than someone prepared to forgo such allegiances in a bureaucratic superstate which can command neither love nor loyalty).

[53] 'My Early Beliefs' and quoted by Noel Annan, in *Our Age* (Fontana, London, 1991), 437.

Orwell's 1941 piece on 'Wells, Hitler and the World State' is worth quoting at length:

Hitler is a criminal lunatic, and Hitler has an army of millions of men, aeroplanes in thousands, tanks in tens of thousands. For his sake a great nation has been willing to overwork itself for six years and then for two years more, whereas for the common-sense, essentially hedonistic world view which Mr Wells puts forward, hardly a human creature is willing to shed a pint of blood. Before you can talk of world reconstruction, or even of peace, you have got to eliminate Hitler, which means bringing into being a dynamic not necessarily the same as that of the Nazis, but probably quite as unacceptable to 'enlightened' and hedonistic people. What has kept England on its feet during the past year? In part, no doubt, some vague idea about a better future, but chiefly the atavistic emotion of patriotism, the ingrained feeling of the English-speaking peoples that they are superior to foreigners. For the last twenty years the main object of English left-wing intellectuals has been to break this feeling down, and, if they had succeeded, we might be watching the S.S. men patrolling the London streets at this moment. . . . The energy that actually shapes the world springs from emotions—racial pride, leader-worship, religious belief, love of war—which liberal intellectuals mechanically write off as anachronisms, and which they have usually destroyed so completely in themselves as to have lost all power of action.[54]

There is clearly a convergence here between Orwell and de Tocqueville, and also, surprisingly perhaps, with Adam Smith. Smith, like de Tocqueville, was an enthusiast for liberalism and the commercial society. But like de Tocqueville, he was aware of its 'inconveniences'. One of the most important of these was that a nation of shopkeepers and consumers, in acquiring a taste for the arts of trade and the luxurious benefits trade makes available would suffer a loss of the martial spirit. This may not have happened as fast as Napoleon (or Smith) might have expected, but there is no denying that the martial virtues are not held in high esteem in Britain or the United States at the end of the twentieth century.

As de Tocqueville observes

trade is the natural enemy of all violent passions. Trade loves moderation, delights in compromise, and is most careful to avoid anger . . . [it] makes men independent of one another and gives them a high idea of their personal importance.[55]

With its emphasis on the individual, it is thus the natural seedbed for democracy, and also the springboard for an insatiable desire for enjoy-

---

[54] Orwell, *Penguin Essays*, 195–6.   [55] de Tocqueville, *Democracy in America*, 637.

ment and material things. Both democracy and a craving for enjoyment militate against the military spirit, which depends on inequality, an instinctive readiness to obey orders, and a certain cultivation of hardship. Democracy militates further against unthought allegiances and prejudices, because each man withdraws more and more into himself, cutting himself off from his fellows and the past. It is in this situation that the overweening welfare state steps in to fill the affective vacuum left by this withdrawal into self, and also to supply the material needs which the democratic individual cannot meet from his own resources. But by then the situation is too late: 'each nation is no more than a flock of timid and hardworking animals with the government as its shepherd'. The shepherd purports to supply material needs 'from cradle to grave', thereby infantilizing those who are dependent on it and whose scope for action is circumscribed by its petty rules and regulations, but is unable of itself to supply any basis for the firm allegiance on which any strong society depends and which makes decisive action, collective and individual, possible. The fact that no such cradle to grave state welfare provision is in fact possible in a modern economy serves only to loosen further the bonds of allegiance the now disillusioned citizens feel for their country, once they realize the impossibility.

Orwell, of course, was arguing that in 1941 things had not gone as far down the democratic road as de Tocqueville suggested, but reflections on de Tocqueville's analysis might suggest to us that Mrs Thatcher's project of combining a strong national myth together with a loosening of the welfare state and an expansion of individual entrepreneurship was not as contradictory as is sometimes supposed. For the same tendency—the democratizing, collectivizing tendency—had both weakened the myth and made entrepreneurship difficult. Strong individuals can flourish in a strong potentially hierarchical society better than they can in one which attempts to deny difference by playing down the ancient myths which both sustain the hierarchies and suggest limits to state action.

Whatever we might conclude about democracy in general, what we need to emphasize in the ethical and political realm is the need for rational reflection on ethical and political values, alongside a realization of the limitations of such reflection. We need to reflect on our values for at least two reasons. First, as human beings we are creatures who rely on judgement in our practical affairs. The situations which daily confront us are multifaceted and subtly different from each other. We are not programmed to deal with them mechanically, nor could we easily envisage programmes to take account of their complexity. So we have to use our reason and intelligence even in knowing how to apply the rules and ideals we have, to determine just what in *this* situation bravery or honesty or kindness requires, or to determine how in this case honesty, say, is to be balanced against kindness.

The exercise of rationality and judgement in the application of particular virtues and ideals cannot be insulated from more global reflection on the virtues and ideals themselves. In thinking about their application, one of the things which will inevitably emerge will be conflicts between values. We will then be induced to decide between them and to rank them, and this in turn can lead to further decisions and discussions over whether, for example, we are to regard, say, manly pride as a virtue at all. We will also be put in positions where we have to decide whether particular virtues or dispositions are to be applied in particular cases, and to judge on the extent of various duties, whether a duty of care to a stricken stranger I meet on the road has to be extended to people I do not meet but about whom I know and whose plight is just as parlous. Or again, does my obligation not to cause pain to human beings extend to sentient but non-human beings, and if so, how far? These are questions of the sort which will inevitably arise in the exercise of any morality, and which will be dealt with by thinkers in any society, however rigid its dogmas and practices. Indeed, if Islam or traditional Catholicism are guides in this area, the more rigid a society's dogmas and practices are, the more need there will be to discuss and define the limits and extent of obligations. What critics deride as casuistry is often the highly developed exercise of rationality and intelligence as applied to the practice of a community which regards itself as settled in its beliefs. But being settled and— again from the point of view of critics—prejudiced, by no means precludes a continuing exercise of rationality and intelligence through which, over time, a society's practices are in fact refined and revised. Such a thing, indeed, is just what would be expected given the nature of moral thinking and the difficulties which arise in practice in the application of any code of principles.

Naturally, too, coming into contact with other moral codes then one's own will spark reflectiveness, but it is important to realize that reflectiveness and rationality are not confined to confrontations between opposing codes. Nor, indeed, are intercultural discussions necessarily the most fruitful route to reflection. They can, on occasion, lead simply to a dogmatic assertion of undefended prejudices or its mirror-image, a weary acquiescence in relativism. What we will then have will be a closing down of reflection and intelligence, an abrogation of our rationality.

Part of the motivation for such an abrogation may actually be a realization that morality and the wisdom needed to make intelligent moral judgements are both founded in the possession of habits, habits which were not adopted initially by individuals because they seemed reasonable to those individuals. The reason for this is not that the habits in question are unreasonable, but that the individuals in question are young children, and, as such in no position to assess their reasonableness. The making of such assessments, indeed, depends on the prior acquisition of the relevant habits, disposing us to see certain courses of action as reasonable and others as unreasonable. Aristotle's thesis might be characterized as maintaining that it is through the courtyard of habit, that we enter the palace of reason.

What Aristotle is insistent on is that questions in morality and politics are not simply questions of cleverness. If they were, we would expect to be able to learn about politics and legislation from Sophists, who profess to teach it but who are far from being able to practise either with any degree of success or wisdom. Would we think we could learn about medicine or painting from people who are not doctors or artists? But politics, too, is a practical skill. The would-be legislator needs experience both of life generally and of politics in particular over and above cleverness and intelligence. It is experience and wisdom which sophists lack, not intelligence.

Practical wisdom, which is necessary for ethics as well as for politics, also needs a training in habits of virtue:

the soul of the student must first have been cultivated by means of habits for noble joy and noble hatred, like earth which is to nourish the seed. For he who lives as passion directs will not hear argument that dissuades him, nor understand it if he does . . . The character, then, must somehow be there already with a kinship to virtue, loving what is noble and hating what is base.[56]

While practical wisdom requires cleverness, cleverness on its own could be harnessed in the service of wickedness. The faculty of cleverness is the ability to hit the goal we set ourselves, but if the goal be bad, then cleverness is mere smartness, not practical wisdom (or phronesis): 'therefore, it is evident that it is impossible to be practically wise without being good'.[57]

If practical reasoning requires right dispositions on the part of those who are going to make good moral judgements, and given that none of

[56] Aristotle, *Nicomachean Ethics* 1179b25–8.   [57] Ibid. 1144b3.

the moral virtues exist in us purely by nature, 'it makes no small difference, then, whether we form habits of one kind or of another from
our very youth: it makes a great difference, or rather *all* the difference'.[58] The point is that we can reason well about value only if we
have been brought up so as to acquire habits of respect for the good
and the virtuous, a respect which assumes both that certain values are
not in any serious sense open to question, and also that the right values
have gradually emerged in the experience and discourse of the many
and the wise.

Aristotle also observes that our powers correspond to our time of
life, and that a particular age brings with it a certain type of intuitive
reason and judgement, and that 'therefore we ought to attend to the
undemonstrated sayings and opinions of experienced and older people
. . . not less than to demonstrations'. Because they have an eye as a
result of their experience, 'they see aright'.[59] Aristotle thus agrees with
Plato that age contributes to moral wisdom, but he disagrees with him
over the significance of demonstrative reasoning.

He disagrees because of a strong sense that demonstrations have
inherent limitations. They take us only so far and, if untempered by
experience, can be misleading. Moral wisdom is concerned primarily
with making the right judgements about the particular matters which
confront us in life, rather than about first principles. If pursued relent-
lessly, with the aim of providing proofs which are of geometric exacti-
tude and persuasiveness, moral argument can cause our moral and
affective life to fall to pieces. According to Aristotle, young men, lacking experience, have no real conviction about moral matters, but only
the language, which they then misunderstand and abuse. And this also
militates against the sense of piety and respect on which moral practice depends. (It may be no coincidence that Plato represents Socrates
undermining the piety of Euthyphro by his negative dialectic, even as
he is himself about to answer charges of impiety. Socrates is often
thought to have the better of a rather brash and unpleasant young man,
but this overlooks the fact that Euthyphro is actually concerned about
a slave who has died as a result of his father's careless attitude to the
life of slaves.)

The obvious problem with the view that practical wisdom is
founded in virtue and that virtue is founded in habit is in the question
of the selection and justification of the relevant habits. Thus, although
nothing said here militates against reasoning about the habits and the

---

[58] *N. Eth.* 1103b24–5.   [59] Ibid. 1143b14.

virtues they subtend, our reasoning will still be constrained by our possession of the relevant habits and dispositions. Is this circle of virtue not argumentatively vicious?

There is no reason why it should be, given that morality is not geometry, analysable in terms of self-evident a priori principles, any more indeed than science is. In both cases, we are *in mediis rebus*. In both cases, too, though in different ways, the *res* in question implies a strong relation to practice and to habit. What we are initially disposed to accept are those beliefs and principles which are already embodied in our practices, and to an extent these beliefs and practices will continue to colour our future reflection, even where we are led to revise them. In the case of morality, there is the further point that morality is closely linked to the affections. How our affections are will have a bearing on what principles we are disposed to accept, and moral reasoning which neglects this fact is likely to remain mere reasoning or to become, in the bad sense, a casuistical attempt to justify practice which falls short of espoused principles.

In stressing the wisdom of experience we are acknowledging what might be called the empirical, experimental nature of our moral discourse: that it is discourse tried and tempered against human passions, with all their recalcitrance and obstreperousness. In recognizing this empirical, experiential element of moral discourse and tradition, we see what is right in Hayek's position. But against Hayek, this understanding does not imply an uncritical attitude to tradition or an irrational attitude to morality; rather it stresses the limitations of criticism and the potentially corrosive effect of rationalism, when that is taken to consist in the relentless searching for reasons for our moral beliefs which Mill, for example, recommended—even where there was in fact no disagreement. Such rationalism can be corrosive because it fails to appreciate that justification in moral matters characteristically issues in fairly short chains of reasoning, and because it may suggest that it is a defect of morality that we can provide no proof that, say, torturing children is wrong which is argumentatively more conclusive than our intuitive conviction that it is (failing to recognize that an analogous situation obtains regarding empirical certainty). Millian rationalism also overlooks the fact that morality is as much a matter of sensibility and practice as of reasoning, and that the presence of morality within a community depends as much on the cultivation of the requisite sensibility in that community as on habits of critical reasoning.

According to Mill in *On Liberty* what is required of us is a systematic scrutiny of even the best founded beliefs and values, even and

especially by the young. On pain of their and our losing 'living apprehension' of a truth, he urged the 'teachers of mankind' to find 'some
contrivance for making the difficulties of a question as present to the
learner's consciousness, as if they were pressed upon him by a dissentient champion, eager for his conversion'[60] *particularly* where there
is general agreement on a truth. One wonders quite what this type of
dialectical questioning is intended to achieve, or likely to achieve
except to convince learners (and teachers) that there is doubt in theory
where none exists in practice, and to make them think that there is
something wrong with a practice which does not admit of explicit
rationalia presentable in geometrical fashion. Certainly, a Millian
approach will give no weight to tradition as such or to any penumbra
of reverence which cleaves to a practice *because* traditional. Nor does
Mill appreciate the extent to which each of us, as a human being, is the
focus of a network of allegiances, rights, and duties, which cannot simply be abandoned until we find rational justification for them, on pain
of violating our identity and affections.

On the other hand, we have Hayek and (in certain moods)
Nietzsche apparently inveighing against anything other than blind following of custom because customary. Nietzsche spoke of that correct
and strong education as being *above all* obedience and custom,
extolling the virtue of subjection to a great guide. Hayek is doubtless
more aware than Mill of the limits of justificationism, and that many
of our beliefs are groundless (and not just in the moral sphere). But, as
we have seen, his apparent irrationalism is unsustainable. Is there,
then, a *via media* between a damaging rationalism and an equally damaging irrationalism, each of which is in its own way oblivious to the
realities of human existence?

Burkean 'Prejudice'

Edmund Burke's famous defence of prejudice provides just such a
middle way, and rather more subtly than is often realized. He writes
that

we are generally men of untaught feelings; that instead of casting away all our
old prejudices, we cherish them to a very considerable degree ... We are afraid
to put men to live and trade each on his own private stock of reason; because
we suspect that this stock in each man is small, and that the individuals would

[60] J. S. Mill, *On Liberty* (Oxford University Press, 1991), 50.

do better to avail themselves of the general bank and capital of nations, and of ages. Many of our men of speculation, instead of exploding general prejudices, employ their sagacity to discover the latent wisdom which prevails in them. If they find what they seek, and they seldom fail, they think it more wise to continue the prejudice, with the reason involved, than to cast away the coat of prejudice, and to leave nothing but the naked reason; because prejudice, with its reason, has a motive to give action to that reason, and an affection which will give it permanence . . . Prejudice renders a man's virtue his habit: and not a series of unconnected acts. Through just prejudice, his duty becomes part of his nature.[61]

Burke, then, argues in favour of a prejudice in favour of old prejudices. This is first because they are old; because they have survived, they are likely to contain some wisdom and fruit of experience. What each of us calls our reason is likely only to be the expression of the unthought prejudices of our age, Millian liberalism, perhaps, or the chimera of world government. So, against our narrow prejudices, we should pose what has emerged unthought over centuries through a certain cunning of reason. We should not disenfranchise the dead, nor disregard what their lives and sufferings have to teach.

Then, secondly, if on examination a general prejudice reveals its rationale to us, we should not simply accept it. We should cherish the institutional form in which the prejudice is clothed, and cherish too its authority over us. If it is a prejudice, it will already have a motivating power, and we should capitalize on that, especially if we are Aristotelians, and appreciate the importance of habit and certainty of decision in the cultivation of virtue.

To draw a contrast between Mill and Burke, we might say this. Where a Millian philosopher would see his role as being the elimination of prejudice and the rationalizing of human affection and allegiance, a Burkean would see a role for a philosopher as an articulator and defender of prejudice; not of any prejudice, to be sure, but as being predisposed to find merit in prejudice, and where merit is found, to reinforce that prejudice.

Such a philosopher would not embark on any quixotic and ultimately self-defeating repudiation of reason, in the manner of Hayek, but he would be far more sensitive to potential and hidden virtues of traditional customs and practices than the rationalist. It will not be a surprise to the Burkean when it emerges that the costs of a Keynesian approach to sexual morality turn out to be high. The difficulty we are

---

[61] E. Burke, *Reflections on the Revolution in France* (Penguin, Harmondsworth, 1986), 133.

now confronted with is that once the prejudice in favour of traditional family life, and against easy divorce and single parenthood has gone, it is very hard to return to it. Even though we are now far more aware than previously of the social costs of divorce, it is almost unimaginable that any government in the Western democracies would act to make divorce harder, at least in part because of the suspicion that a harder law would simply be ignored by those who have lost their prejudice in favour of the sanctity of marriage. In the autumn of 1993 in Britain we thus had the paradoxical spectacle of a government supposedly committed to traditional moral values under a Lord Chancellor renowned for the strictness of his own morals actually proposing a law to make divorce easier.

What I am intending to suggest here is that morality is first and foremost a practice, one which stems from other practices and traditions within a community, and which is ultimately based in facts about our biological nature. What is being advocated here is not relativism or irrationalism. It is not being said that we do not or should not reflect on or revise our values. Nor, against Ruse on the one hand, and Hayek on the other, is it being said that biology and tradition constrain our behaviour in such a way as to make reform or moral progress impossible. Part of my argument earlier was to insist that morality, however narrow and circumscribed, goes beyond the fundamentally selfish impulses of which Darwinian biology treats. Moral progress, in so far as it exists, consists in deepening one's understanding of what mutual recognition involves, and also in appreciating how the extent of one's obligations may increase. On both counts, early Christian advocacy of charity to any needy person, irrespective of status or nationality represented progress. By the same token, Nietzschean contempt for the rabble represents a retrograde step, with consequences all too real.

What, though, must be borne in mind are the dangers inherent in departing from the existing moral practice of a people, particularly if this is done in the name of some abstract principle which is held to override all or any local obligations. An example of what I have in mind is that provided by George Orwell in his discussion of world government. Talk of the rights of man or of universal equality can easily be used to suggest that we have no particular obligations to our own people or region, and that such local or national associations as exist should be dissolved in favour of supranational governments under which local arrangements are replaced by universal homogeneity. This attitude is fairly widespread among those of what Orwell calls 'ordinary education', as we have seen. As the writer Alan Bennett once

put it, 'Not English I feel now. No country. This is just where I happen to have been put down. No party. No church. No voice.'

It is, of course, true that there is much in any tradition, whether it
be an aspect of customary morality or a local association, which has
not been planned, and which has emerged or developed gradually,
haphazardly even. The rationalist tendency is to dislike unplanned and
haphazard institutions, and to prefer easily statable principles and
clean lines of development. Although the position I am adumbrating
is not mindless conservatism—it embraces both development and
reflection—there is a clear difference of emphasis between it and rationalism. This difference can be outlined under three heads.

First, a point already alluded to in connection with divorce, the cost
of doing away with custom and tradition can be rather higher than will
be apparent to the rationalist mind. For the rationalist, a husband and
a wife are a man and a woman who enter a contract for a number of
purposes. If the purposes are not being fulfilled, and the contract is
tiresome to the parties, they may agree to tear it up. But the downgrading of marriage from sacred obligation to defeasible contract has
not in any obvious sense improved society. There may have been wisdom in the old arrangement, at least for the vast majority of the population: one can hardly view the increase in single parenthood from 8
per cent to 33 per cent in 50 years as unalloyed social progress, given
that the costs and difficulties of this arrangement are becoming ever
more obvious.

But, secondly, it is not simply that the customary and the traditional
may embody wisdom of a consequential sort, which the reductive tendency of rationalism is likely to overlook. There is further the way in
which the language and concepts of tradition provide what Burke
refers to as 'pleasing illusions' and 'the decent drapery of life', which
make 'power gentle and obedience liberal', harmonizing the different
shades of life. By contrast, the rationalist would reduce all this to the
elements beneath, which all have in common and which can be
described without appeal to historically contingent differentia:

On this [reductive] scheme of things, a king is but a man: a queen is but a
woman . . . Regicide, and parricide and sacrilege, are but fictions of supersti-
tion, corrupting jurisprudence by destroying its simplicity. The murder of a
king, or a queen, or a bishop, or a father, are only common homicide.[62]

Burke goes on to speak of this as a cold and barbarous philosophy,
devoid of taste and elegance, but not of taste and elegance only. In the

[62] Burke, *Reflections on the Revolution in France*, 171.

reductive, rationalizing scheme of things, laws will be obeyed only through a combination of terror and self-interest: 'nothing is left which engages the affections on the part of the commonwealth'. In their different ways, centralized socialist planning, as practised particularly but not solely in Eastern Europe and free market individualism have begun to erode the purely affective, the ceremonial aspects of life. Burke's point would be that in so doing, in so far as they have been successful, they have not increased cohesiveness or community. They are likely only to make what Alan Bennett claims to be the case for him a reality for us all. It has yet to be seen whether the abstractions and bureaucracies of rationalist politics can command anything approaching the lively allegiances of nation states and their institutions.

A third way in which rationalist approaches to morality and politics differ from traditionalist ones is in their respective approach to principles and goals. The rationalist characteristically looks at institutions and customs in terms of the principles or goals he sees them as serving. Does, say, the bar on lying serve greater happiness overall? Is affirmative action consistent with the principle of treating everyone with equal respect? Does a particular inequality of distribution nevertheless increase the well-being of the worst-off, and so remain consistent with fairness, abstractly conceived? The suggestion in each of these rather different cases is that it is possible to take a practice or policy and judge it against some overriding set of principles, whose meaning and justification stand independently of their particular concrete applications.

Against this, it is necessary to stress that notions like happiness, respect, and fairness cannot be understood outside the particular contexts in which particular people would recognize them. They are not, as it were, transcendentals against which all particular societies can be judged, for in judging them one will inevitably be importing one's own conceptions of happiness, respect, and fairness derived from one's way of life. They cannot, therefore, be regarded as goals to be approached indifferently by various means. Indeed, what one is prepared to consider as means to, say, happiness or fairness says a lot about what one considers happiness, or fairness to be. For some, for example, it would be fair for the state to remove someone else's lawfully acquired property in order to redistribute it; for others this would be the height of unfairness. Similarly, over happiness there are equally intractable discussions about the value of creative discontent, and whether such a thing would be worth jeopardizing a settled, but mediocre, existence for.

If, then, one is to enunciate a principle—such as strive to be fair in one's dealings—what the principle means only becomes clear by seeing the practices which one would take as exemplifying fairness. A Rawlsian would have very different examples in mind from an Aristotelian. In this sense, principles are essentially abridgements of practice, and carry the weight they do for particular groups of people because of their rootedness in the practices those people actually accept and find compelling.

Rationalists sometimes think that they can convince people to extend or radically revise their moral practices simply by appeal to principle. For example, one sometimes hears it argued that because we have a primary, impartial, and pressing obligation to relieve serious suffering before we engage in acts of enjoyable but unnecessary generosity to those around us, we should spend all our surplus income (which is what?) on the starving and needy in remote parts of the world rather than on our own families. An immediate reaction to this is to make the Aristotelian point that such arguments are rarely put by those with much experience of life or with much insight into human affection. But the point is not just that moral argument should respect the empirical extent of human affection and sense of duty, though it is that. It is further that our understanding of the relief of suffering principle is paradigmatically based in cases where the suffering in question is close to us and can be directly relieved by our own actions. Like many things in morality, it is just not clear how far this principle is to be extended beyond the cases in which it is incontrovertibly compelling.

Within a particular tradition or conception of life, there may well be general agreement over which problems need urgent attention and what the means are which might be acceptable. Thus, for example, there is widespread agreement that it is undesirable that people should sleep on the streets in London or New York; governments are under pressure to take action to prevent this and many people might well be prepared to contribute some taxes to solving the problem. But what about the far worse situations in Calcutta or Buenos Aires? People are still people in these places, with, it seems, as much right to be treated with respect and concern, and doing something effective for Third World beggars would (as aid agencies constantly tell us) cost very little by Western standards. But, whatever the abstract arguments philosophers and others might mount, it is perfectly clear that the majority of people in the West are not prepared to compromise their living standards substantially to help people in India or Brazil. They do, though, consent more or less readily to having considerable proportions of their income devoted to the provision of welfare in their own countries, and not just, one feels, because they think that they might one day rely on welfare themselves.

The Limits of Obligation

Leaving aside questions as to the effectiveness of welfare, either at home or abroad, the situation seems to be this. In this, as in other areas, individuals understand their moral and political obligations through the habits and dispositions they acquire in their upbringing. These habits and dispositions tend naturally to be those current in the society or sub-group to which they belong, as a result of which they find themselves cued into discussions and practices with people similarly brought up. All share intimations as to the limits and extent of their duties, rights, and obligations, with agreement in central cases (both of obligations and of non-obligations) and creative disagreements at the margins. These disagreements are not necessarily resolvable. Is, for example, a foetus of twenty weeks entitled to the same care and concern as a new-born baby? Should the state legalize soft drugs? Is free health care for all a desirable political goal?

But these and other well-known areas of dispute are marginal, in the sense that they are discussed with reference to central cases on which there is general agreement and in terms which command general agreement. It is certainly imaginable that on any of these issues in our societies a broad consensus and a corresponding shift of opinion could occur, so that in a hundred years these now vexed issues were not issues at all.

What, though, is far more difficult to imagine is a situation in which, say, everyone in the world is seen as part of the social contract in which we all admit pressing duties and obligations to each other as we do within our national boundaries.

Humanity as a whole does not command the same respect as shared nationality and citizenship: geographical and psychological and cultural and religious proximity do in practice constrain people's sense of mutual obligation. Those who would wish to weaken the effect of such proximities ought to reflect that in doing so they may simply destroy whatever sense of mutual obligation does exist, without putting any very strong sense of community in its place. And there is the further point that a moral system which imposes obligations which people cannot or will not realistically discharge may have the unintended effect of bringing morality as a whole into disrepute.

But this does not necessarily show that it is right that the extent of our moral obligations should be confined to the historically contingent boundaries in which our sense of moral obligation is first realized, and to which, arguably it was tailored. Our earlier reflections on the gaze and on self-consciousness did not say that one should respond only to the gaze of those close to one or those who can benefit one. Morality was distinguished from prudence in precisely this way. In a sense nowadays we do gaze on and are gazed at by people very remote from us.

What we may have here is a situation analogous to that which obtains in epistemology. We inherit certain dispositions, such as perceptual apparatus, sympathy to those around us, traditions of thought and behaviour. All these have developed in particular circumstances, and in response to specific and restricted conditions (the human environment, life in comparatively small and cohesive communities). But we also reflect on these inheritances in ways which break down the initial limits of their applicability. The gaze is originally a very personal, very individual thing, a particular psychological event. But its ramifications go far beyond those I personally and directly encounter, and, as with science, lead us into areas where our intuitions and principles are hard to apply. In the closing pages of this chapter, I have argued that we should be cautious of undermining the moral traditions we have, and that we should see our principles as receiving their first definition and justification within our practices. But advocating a decent degree of conservatism in practice does not imply that reflection on our practices in changed circumstances may not bring us up against the limits both of our principles and our nature.

In this sense there may be something inherently utopian, and in the real world even tragic about our moral sense. It is practicable and workable and fully motivating only within limited environments, particular communities, and well-understood circumstances. On the other hand, once we reason about our morality, we see the contingency of the circumstances and the universality of the reasons. The situation is further destabilized by the development of new techniques which radically alter our understanding of both life and death, to say nothing of the vastly increased numbers and concentrations of human beings in the world. Principles which were at one time comparatively straightforward to apply become frighteningly unclear with increases in knowledge and power.

A restriction of moral sensibility and moral concern to historical limits is bound to seem arbitrary and irrational, as indeed it will be. We can neither justify an attitude of complete indifference to the sufferings of others, however distant in space or culture, nor can such an indifference be maintained without a coarsening of our moral sensibility to those closer at home. On the other hand, if a stress on principle leads, as it may, to a weakening of all sense of local allegiance the danger is that any sense of moral motivation will evaporate. The dilemmas which result from the tension between reason and tradition in the moral sphere may be soluble; but there is no guarantee that they will be, and to play down the role in moral practice of either reason or tradition will put morality itself at risk.

7

Beauty and the Theory of Evolution

The Natural History of Beauty from Flint Axes to Russian Icons

We are constantly reminded of the small amount of time it has taken for human beings to move from the Stone Age to modernity. In some ways a certain agnosticism seems preferable here to the gee whizzes of popular science. How long should it take an intelligent tool-making species to move from the wheel to the jet engine? Is it really so very amazing that while bacterial life started on earth more than three billion years ago, land plants apparently date from a mere 400 million years ago, shortly after the first fish and about the same time as the first amphibia? What, in any case, is the significance of talk of vast tracts of time in the absence of creatures who can plan, reflect on what they are doing, and come to an awareness of the passing of time, and hence of tasks accomplished fast or slowly?

What is more significant than the mere passage of time, whether this is long or short, are the transitions in emphasis in human culture from activities which are directed to survival and reproduction to those which are not. Hunting and gathering, eating, sleeping, reproducing and caring for the young are activities without which neither individuals nor species would survive. They also take up an awful lot of time, particularly if you do not know how to cook, plant seeds, domesticate animals, or organize your economy on the basis of specialization and the division of labour. If you are having to spend most of your day simply surviving and providing for your dependants, you are not going to find much time for developing the higher activities of art, science, morality, and religion, even if (as I believe is the case) there are in reflective, self-conscious beings drives in each of these directions, however simple or sophisticated their economies.

By around 10,000 BC mankind had learned to farm and in a few places was beginning to live in towns. The great period of prehistoric enslavement to need was coming to an end. Much time was saved for at least some specialists, and civilization could begin its ever more intense progression away from the satisfaction of basic drives. There had, of course, been technology of a sort and art of a sort long before that. Tool and weapon using and making had been a part of human and pre-human life for 1,000,000 years or more. Art is more difficult to discern, but we know of humanly fashioned hand axes of 250,000 years ago which *look* beautiful to us, of sculptures, presumed to be cult objects, from 30,000 to 25,000 BC, and of drawings and carvings from a couple of thousand years later, and certainly from before 20,000 BC.

There has been much speculation about the significance of prehistoric art objects, and about 'beautiful' axes. Was the maker of the Solutrian 'laurel leaf' flint axe in the Musée de l'Homme in Paris guided by conscious considerations of beauty? Or are its delicate patterns and symmetry due to some purely functional role they play in the activity of cutting? Or are they due to some genetically based but unconscious preference on the part of its maker? We shall, of course, never know for certain. What, though, is evident even at this stage of our inquiry is that an apprehension of beauty assumes on the part of the perceiver or maker a sense that he is admiring or designing the thing in question *because* it has or he wants to give it a pleasing form. It is more than just a going for something which may incidentally please others on instinct or for purely pragmatic considerations.

Similar problems attend the evaluation of prehistoric cult objects. Are they simply magical tokens, designed to placate a god or to achieve other more basic purposes to do with fertility? Or are they, in addition, manifestations of a genuinely aesthetic sensibility? It is interesting here to compare the Willendorf Venus, which seems crude and unaesthetic even in comparison to contemporaneous tools and weapons, with the haematite torso of about the same period from Ostrava Petrkovice (now in the Brno Archaeological Institute), which prefigures the proportions and even the stance of a late classical or Hellenistic Venus.[1] It is very tempting to think that there was a fine aesthetic sensibility at work in Moravia 27,000 years ago, if only because what we have in the torso is clearly worked, and worked in a way which emphasizes what for us is the very paradigm of the aesthetic. Would it not be too great a coincidence if some prehistoric Moravian cult required attention to just what are for us aesthetic characteristics in its cult objects for purely cultic reasons, as opposed to

---

[1] Cf. N. K. Sanders, *Prehistoric Art in Europe* (Penguin, Harmondsworth, 1968), 11.

their being put there because in part at least the carver found them pleasing?

Whatever might be said about the aesthetics of prehistoric art and artefacts, though, what is as striking as anything in human history is the fact that by AD 1400 in technologically and culturally backward Russia an art of extreme technical competence had developed, whose explicit intention was to present human beings not as hunters or gatherers, not as survivors or reproducers, not as physically beautiful or as technically competent but as divinely transfigured, owing allegiance only to a vision of eternity which transcended all material considerations. What is striking and suggestive about this is the way it suggests that once a degree of freedom from immediate want is gained in human affairs non-utilitarian motivation can begin to play a dominant role.

For icons, though certainly works of art, in that they are painted with an eye to aesthetic virtue, use the aesthetic purity of the image to witness to an image of man as destined for and capable of achieving a mode of life quite different from that of the nature of which biologists and anthropologists speak. I am not here assessing the truth or falsity of Orthodoxy, but simply pointing to the fact of its existence and to its iconography as showing how human beings are capable of being motivated by a striving beyond the instinctive. Nor, indeed, would it seem to have much to do with any 'tournament of the mind' aimed at success and admiration of a worldly sort if only because the whole motivation of icon painting was a sublimation of worldly instinct. That we could go from ape to caveman to icon painter: now that is a truly remarkable transition, however fast or slow, and even though the transition may be exploiting continuities already hinted at.

Continuities

Whatever else we may say about beauty, we must not forget that people are attracted by beauty. Appreciating something beautiful makes us want to be with it. Beauty gives pleasure by drawing on the affective dimensions of consciousness.

Consciousness, as we have seen, is not purely a matter of information processing: along with information processing, in consciousness we also become aware of the pleasures and pains associated with things and their appearance. Further, we take delight in appearances themselves, in experiences of beauty. And this leads people, not unnaturally, to see our appreciation of beauty as being rooted in instinct and in our neural circuitry.

Aesthetic contrivances, writes E. O. Wilson, 'play upon the circuitry of the brain's limbic system in a way that ultimately promotes survival and reproduction'.[2] Wilson's idea is that aesthetic interest has been embedded in us because of the usefulness of curiosity and the search for connections and similarities. For Darwin, aesthetics is more directly linked to reproduction, but he would not have demurred at Wilson's efforts to ground it in our genes.

In *The Descent of Man* Darwin had argued that the elaborate displays found among animals and birds, particularly during mating, could be explained only on the assumption that birds and animals were capable of appreciating beauty. One could not otherwise explain why it was that so much effort went into these things, though he added that 'why certain bright colours should excite pleasure cannot be explained', beyond saying that habit and an in-built preference for regularity and symmetry come into play.[3]

Actually the position is not nearly as straightforward as Darwin thought, and the whole question has evoked controversy ever since. In *The Descent of Man* Darwin was concerned to demonstrate continuity between humans and other animals. If he could show that there existed in animals a genuine aesthetic sense, then one of the major ostensible differences between humans and animals would have been eliminated. The tails of male peacocks and the songs of many male birds are, in Darwin's terms, elaborate, graceful, splendid, beautiful, and highly ornamented. That the females of the species appreciate these things—just as human females do, by decking themselves with the plumes of birds—cannot be denied, says Darwin. It is the most splendid males which succeed in getting females, and this, at any rate in the case of peacocks, has been confirmed since Darwin by experiment and observation.[4]

What has not been confirmed, though, is whether the females who choose the splendid peacocks do so for aesthetic reasons, pure or impure. This, indeed, was the subject of an early dispute between Darwin and Alfred Russel Wallace. Wallace was, in general, inclined to emphasize the differences between humans and (other) animals, a

---

[2] E. O. Wilson, *Biophilia* (Harvard University Press, 1984), 61

[3] Charles Darwin, *The Descent of Man*, i. 141.

[4] Cf. Helena Cronin, *The Ant and the Peacock* (Cambridge University Press, 1993), 225, where Cronin refers to empirical work by Marion Petrie. My analysis of evolutionary accounts of beauty draws heavily on Cronin, though my conclusions are quite different from hers.

point to which we will return. But specifically in the matter of female choice, Wallace was far more selectionist than Darwin. For Wallace, female choice could not be purely capricious, at least not if it were to play a significant role in the development of a species. Darwin had, of course, been arguing that we can explain the ever more splendid, ever more elaborate tails and songs observed in the kingdom of the birds by selection pressures. The most splendid males are chosen by the females, who, in turn produce splendid males. The most splendid of the new males are then chosen in their turn by the ever more choosy females born of the original choosers, and one more twist is given to the spiral and so it will go on until checked by some countervailing pressure. The evolutionary possibility of this sort of selection pressure and its trajectory was well demonstrated by R. A. Fisher in 1930. Fisher showed that once a selection preference gets embedded in a population, it moves forward with ever increasing rapidity under its own momentum, even where (as in the case of the unwieldiness of the peacock's tail) some other disadvantages accrue to the beautiful and favoured males as a result (which does, of course, weaken Darwin's claim about the speedy eradication of injurious variations).[5]

While Darwin, supplemented by Fisher, shows that in a particular species the development of (to us) ever more aesthetically striking characteristics can happen as a result of female choice, it still does not show that the females are choosing for aesthetic reasons. That they might not be is suggested by the possibility that beauty is linked to other more clearly adaptive properties. This indeed was how the selectionist Wallace argued, his selectionism here going hand in hand with his claim that *with us* aesthetic response is part of our 'spiritual nature' and cannot be explained on 'purely utilitarian principles'. Wallace proposed that the apparently aesthetic choices of peahens were in fact 'sensible', that is the (*to us*) aesthetically pleasing properties of peacocks, humming birds, and the like were linked to more directly advantageous traits.[6]

For example, a beautiful appearance might be a sign of good health, or of good genes more generally. Even where beauty seems to be to a creature's disadvantage, the fact that it is able to overcome the disadvantage may indicate its greater inner strength. Ornamental display may be a sign of territorial dominance, and so indicating general strength and superiority. And so a gene might develop in peahens for

[5] Cf. R. A. Fisher, *The Genetical Theory of Natural Selection* (Clarendon Press, Oxford, 1930).
[6] On Wallace, cf. Cronin, *The Ant and the Peacock*, 186–91.

preference for peacocks with beautiful tails, not because their tails are beautiful and even less because peahens consciously appreciated their beauty, but simply because peahens choosing those peacocks by virtue of their tails (though not by virtue of their tails' beauty) would tend to have similar offspring.

How might it be possible to decide between the view that aesthetic preference in animals is purely aesthetic or that what looks like aesthetic preference to us is really a preference for other qualities which go along with the aesthetic and are not chosen *via* any aesthetic appreciation? The fact of the matter is that there is no way of deciding, because all we have is broad-brush behaviour which is capable of different interpretations. As observed in Chapter 3, it is hard to attribute specific beliefs or intentions to non-linguistic creatures, because their behaviour is consistent with many meanings. When the female chooses her peacock with the most beautiful tail, it is certainly as if she was choosing him for his beauty. But it might equally be that she was genetically programmed to respond to the tail with the highest number of eye-spots, with appreciation of beauty coming into it not at all. Even if, against Wallace, the magnificent tail confers or is associated with no more general advantage to its owner, as Fisher has shown, only a little and initially inexplicable disposition on the part of some females would be needed to get the whole history of peacocks' tails under way. Any female preference for a trait such as well-tailed peacocks, if widely distributed in a population, could lead to ever more well-endowed peacocks as being the ones both chosen in themselves and more likely to produce further offspring of the same sort, who would in their turn manifest the same characteristics and preferences. And it need not be the case either at the start or subsequently that the female choice was either aesthetic or sensible on other grounds. It would be enough for it just to be there and not too harmful.

Indeed, if, as I think we should, we follow Kant's analysis of the aesthetic judgement,[7] I am rather inclined to think the peahen's preference could not be genuinely aesthetic. For Kant, the aesthetic judgement is disinterested, universal, and non-cognitive. I shall have more to say shortly about the supposed non-cognitive nature of the aesthetic judgement, but I want first to concentrate on its disinterestedness and universality, because it is clear that whatever else female peacocks are doing in going for the best-tailed male, they are not making a disinterested or universal judgement. By saying, first, that

[7] In Kant's *Critique of Judgement* (1780), sects. 1–60.

the aesthetic judgement (or what Kant calls the judgement of taste) is disinterested, Kant means that we are not interested in consuming or processing the object we are reacting to. We are interested in the 'real existence' of the object, and contemplate it for its own sake, and not as something we could consume or sell or use.

Hence Kant would distinguish clearly between those elements in an art dealer's evaluation of a work of art which pertain to a work's saleability, and those which pertain to its aesthetic qualities. Similarly, we could imagine a landowner appreciating his estate for both its beauty and for its contribution to his reputation. The disinterestedness of the aesthetic judgement can be characterized in terms of the object in question being contemplated rather than used, and this goes even for the case where one admires something for its fittedness to function. That is to say, we can speak of a beautiful car or a beautiful aeroplane, and part of what we are admiring would be the way its design refers to what it is intended to do, and in some way symbolizes that. But in saying that the car is a beautiful car, we will be standing back from using it, contemplating its line or design without actually using it, even if the contemplation eventually issues in use. Aesthetic enjoyment is, in Kant's terms, a 'reflective' rather than an 'organic' or 'appetitive' pleasure.

Matters become more complicated when we turn to sexual beauty. In admiring a beautiful woman is the man disinterested? Or is he looking at her as a potential or even as an actual lover? This, as we shall see, was a question which greatly taxed Plato. The following, though, is clear. There can be a chaste, non-possessive appreciation of female beauty, either in the flesh or in a painted image. *Pace* John Berger, not every painted nude is an invitation or incitement to male possession or fantasy, even if many are. Titian's *Education of Love* (Borghese Gallery, Rome) is certainly a celebration of female flesh, whose artistic triumph is that it is at the same time a reminder of the poignancy of love and the pain of the experience (in which Venus is binding the love-inducing gaze of Cupid). Equally, and despite what I once heard a gallery lecturer say, Titian's Diana in his *Death of Actaeon* (National Gallery, London) is not there as an excuse to paint and admire a strapping wench, but rather as an evocation of that strain of heartless and perhaps purifying beauty which is so marked a feature of the Olympian pantheon. (Diana and Apollo are sister and brother and it is Apollo who presides over *The Flaying of Marsyas*, the other great allegorical masterpiece from Titian's later years.) Even the Venus of Urbino and the Rokeby Venus, though certainly reminders of the most organic pleasures possible, are reflective reminders, which can be viewed aesthetically, that is disinterestedly and reflectively. And if we—even males—can look at these works of Titian and Velázquez, and perceive them as beautiful, without desire being engaged, is it not *possible* for us—for men, even—to admire reflectively, undesiringly, the beauty of a beautiful women or boy? I say boy here because we have in Plato's *Symposium* (219a–d) a powerful image of just this phenomenon. Socrates, who loved beauty, and who loved Alcibiades' beauty, was nevertheless able to spend a night in Alcibiades' arms, yet chastely.

What, though, is striking about the reflectiveness which Kant argues is essential for a true aesthetic judgement is that there is no obvious analogue in the animal realm. We say that the peahen admires the cock's tail only because she chooses the cock. The judgement, if such it is, is interested, organic: the manifestation of a very basic desire is, indeed, the only criterion of its existence. We hear no stories of animals reflectively, disinterestedly admiring some aesthetic feature of their environment. Even if, as is not the case, we found animals simply gazing at a beautiful sunrise, say, we would not be able to establish that it was an aesthetic judgement. How could we, indeed, when the animal's action is the only basis on which we have to attribute any judgement or belief at all, and action alone would never be finely nuanced enough to distinguish between an aesthetic choice and a utilitarian one? But this being so, we are unable to attribute a genuinely aesthetic judgement to any non-linguistic creature, for behaviour (or lack of it) is too coarse a discriminator to allow us to tell whether or not a creature is disinterestedly admiring some property in the world around it.

Much the same applies to Kant's second mark of the aesthetic judgement: its universality. When the peahen chooses the peacock with the most luxuriant tail, she is not implying that all peahens ought to concur in finding it beautiful. Nor, when I declare a preference for butterscotch over toffee am I saying that every right-minded person must agree. When, though, I compare the taste of mature Burgundy with this year's *vin de table*, I am beginning to express something like a true judgement (Kant's rather dismissive remarks on Canary wine notwithstanding); that is, I am saying that everyone ought to judge the same way as me. And certainly when I say that Beethoven's late quartets are great works of art, spiritually deep, full of exquisite lyricism, earthy humour, and fascinating counterpoint, I am intending that what I say should command general assent.

The fact that many of Beethoven's contemporaries failed to appreciate the late quartets means that I am committed to saying that they were in error over them: that there was some failure in their perception. I am not simply saying that Beethoven's contemporaries were unable to see that future generations would find the quartets deep, humorous, lyrical, and so on.

The most natural way of explaining the Kantian universality of the aesthetic judgement—to which, I stress, no equivalent can be found in the kingdom of non-speaking animals—is to say that those who fail to agree with a generally acceptable aesthetic judgement are misperceiving something in the work in question. Thus, Beethoven's contemporaries, in my view, though not in Kant's, misperceived the works, misreacted to their qualities, because of their surface strangeness, or for some other reason.

There are two reasons why Kant does not wish to explain the intended universality of my aesthetic judgement in terms of objective properties of the aesthetic object. Both are to do with the way an aesthetic judgement involves the feelings of the perceiver. First, aesthetic judgements are characteristically singular; that is, they are singular judgements about single objects. To make such a judgement requires that the person making the judgement is acquainted with the object or objects in question. I can judge that, say, the Rokeby Venus is a beautiful painting only if I am perceiving *it* or have perceived *it*. An account of the painting without the experience will not be sufficient. Even if I were able, dubiously, to list all its properties there would be no guarantee that in some other context even only slightly different from that produced by Velázquez just those properties might not produce a quite different and unacceptable result.

Actually, I suspect that this extreme sensitivity to context does not apply in the peahen or other animal cases. In a series of observations of peahen–peacock matings, in ten out of eleven successful matings the female chose the male with the highest number of eye-spots in the tail, and in the odd case the chosen male had only one spot less. This doesn't look much like Ruskin explaining why in Veronese's *Presentation of the Queen of Sheba* (Galleria Sabauda, Turin), what would in other hands have been a mass of 'trivial or even ludicrous detail' here detracts in no way from the nobleness of the whole.[8] The animal cases, by contrast, look much more like genetically programmed responses to general and easily generalizable features of the environment, requiring experience of the feature or features in question, but hardly reflective concentration on a specific object.

---

[8] John Ruskin, *Modern Painters*, vol. v (1860), part IX, sect. II, pp. 293–4.

That the aesthetic judgement is singular and not generalizable does not imply by itself that it is not objective, nor that it is not about real properties of objects. Nevertheless we might well speculate that this emphasis on the particular and on its singular nature, and on its irreducibility to rule or to formal algorithm has been a strong factor in getting people, perhaps even Kant himself, to think of aesthetic judgements as fundamentally subjective, as Kant does. Science and its methods have such a grip on our imagination that, despite all the evidence we have to the contrary from daily life and from the everyday use of language, we find it hard to envisage that a judgement could be at the same time not formalizable *and* objective.

Kant did see the aesthetic judgement as non-objective: that is he saw calling something beautiful as referring not to qualities of the object to which we are applying various genuine concepts (lyrical, deep, humorous), but rather to the reactions we have to the object. The determining ground of the judgement of taste he says, right at the start of the *Critique of Judgement* (in section 1) 'can be no other than subjective'. For Kant the judgement of beauty is not a cognitive judgement; it refers rather to the feelings of pleasure or displeasure we have in the object in question, and we determine its beauty or lack of beauty by referring it in imagination to the feelings it provokes in ourselves and which we think it *ought* to provoke in others (not, note, that it necessarily will). Thinking of the aesthetic judgement as normative in this way again distinguishes it from the purely behavioural response of the peahen to the most spotted tail.

Kant's assertion of the subjectivity of the aesthetic judgement is not simply due to its unformalizability, which in any case would be an inconclusive reason. It is due more to his realization of what might be referred to as the 'magnetism' of value: that is to say, if beauty is simply a property of an object, like its squareness or redness, how is it that recognizing beauty in an object *ipso facto* attracts us to it (cf. 'virtuous', 'wicked')? If the recognition of beauty or of wickedness were due to something happening in us, then *that* problem would be quickly solved. But, I shall now argue, it would be solved far too quickly.

In the first place, it is not true even that *every* time I recognize an object as having specific aesthetic or moral qualities, I am or become personally involved in the appropriate way. For a long time I was well aware of the aesthetic fineness of Raphael's work, of its beauty, without ever being particularly attracted to it (partly because of its religious associations), and Bach's wonderful music—which, again, I did perceive as such—also often left me cold. The point is that while aesthetic and moral properties do characteristically engage us, and are doubtless rooted in our genetically given constitution, they can and do transcend their beginnings in such a way as to open at least a theoretical gap between recognition of property and human response in particular cases. No such normative gap—a gap allowing us to say someone *ought* to like or admire some aesthetic feature of an object—exists in the animal case, where animals either respond or fail to respond. There would just be no sense in arguing that a peahen who failed to respond to luxuriant tails was making a *mistake* in judgement, as opposed to lacking a particular behavioural response.

In analysing aesthetic judgements as obliquely referring to the emotions which objects of various sorts stir up in us, Kant was, as in so much else, following Hume. For Hume, seeing an object as beautiful or ugly is a matter of 'gilding or staining it with the colours borrowed from internal sentiment'. Nevertheless, like Kant, Hume is also insistent on the universality of the aesthetic:

the same Homer, who pleased at Athens and Rome two thousand years ago, is still admired at Paris and at London. All the changes of climate, government, religion, and language have not been able to obscure his glory. Authority or prejudice may be able to give a temporary vogue to a bad poet or orator; but his reputation will never be durable or general. When his compositions are examined by posterity or foreigners, the enchantment is dissipated and his faults appear in their true colours. On the contrary, a real genius, the longer his works endure, and the more wide they are spread, the more sincere is the admiration which they meet with.[9]

Hume, like Kant, wants to hold *both* that aesthetic judgements are potentially universal *and* that their source and ultimate reference is human taste; *both* that there is such a thing as good taste, vindicated and manifested through the test of time, *and* that this good taste is not warranted by the objects themselves, but is a matter of those objects attracting our feelings or (in Hume's case) being gilded and stained by them, and exists merely in our minds. For Hume, as for Kant, the cross-cultural and trans-temporal critical-creative dialogue which elicits the cool, reflective, and disinterested judgement, say, that Homer is a poet of great stature, is ultimately a dialogue about what our feelings ought to be, not about anything objectively or really there.

But why 'ought to be' and how universal *if* these judgements aren't based in something actually there, and actually there apart from any

---

[9] David Hume, 'Of the Standard of Taste' (1757), para. 11.

feelings I or anyone else happen to have at a given time? And what is our dialogue *about* if it is not a dialogue about the *Iliad* itself? Because I find the attempt in Hume and Kant to subjectivize the aesthetic judgement hard to accept in itself, and particularly hard to accept in the context of its supposed normativity and universality, I now want to suggest an approach to beauty which, while objective, does not deny either its roots in human, and ultimately instinctual responses or that it is a valuational—that is, magnetizing—concept.

Objectivity of Beauty

On the view I am putting forward, to find the Rokeby Venus beautiful is to make an objective judgement (and one which most people acquainted with the painting would certainly agree with), but it is not to make a judgement which is factual in the scientific sense. That is to say, it is a judgement which needs a specifically human sensibility to discern or discuss. It is not a judgement about a measurable fact, or about anything whose existence can be uncovered by the application of a rule. Against Hume and Kant, and in line with what I argued in Chapter 5 about secondary qualities, I want to urge that these may be disclosures of reality which can become apparent only to perceivers or agents constituted in specific ways, with particular structures of feeling and sensibility, and which are not based on the application of a rule or rules.

As became clear in considering secondary qualities such as colours, sounds, and smells, just because certain features of the world became apparent only to a particular type of sensibility, it does not follow that what is perceived is not real or objective, or cannot be a valid disclosure of reality. Each type of case must be judged on its merits. To assert that no such case can reveal anything real or objective is to make the questionable and question-begging assumption that natural science is the touchstone of reality and that only the primary qualities it reveals are truly objective. Once that is done, of course, secondary qualities, morality, aesthetic properties, and everything else outside of natural science, including perhaps consciousness itself and its contents become illusions or illusion-generating, at best misleading epiphenomena riding on the back of the causally more basic primary qualities or in the some other way 'supervening' on them.

Even though there can be no knock-down argument against the foundational nature of natural science, except an appeal to experience, we should not forget that in the end science itself comes to be treated as fact rather than fiction only because it is confirmed in experience. In the end, it is only through their conformity to, and explanatory power over the experiences we have that we grant the statements of natural science the accolade objective. And, in its own realm, an objectivist account of our aesthetic judgements explains our experience of beauty, ugliness, and other aesthetic properties in a way the subjectivist accounts of Hume and Kant cannot.

What is meant by talking of an objectivist account of aesthetic value is this: while our judgements in this area are *our* judgements, judgements made possible by our biological constitution and cultural traditions, they none the less reveal genuine properties of the objects in question. When Hume and his followers, including here Kant, say that in finding a sunset or a painting beautiful, they are 'gilding and staining' the world from within their own breast, or that in so judging they are not applying any true concept to the object so described, they are in fact saying little more than that beauty is not a formalizable property, nor one which features in certain types of scientific theory. To be sure, as is already hinted at by the difficulty of finding rules for beauty, and by its extreme context dependence, there will be no set of physically determinable properties shared by and only by beautiful objects (sunsets, sunflowers, pictures of sunflowers, musical evocations of sunsets). But that a Mediterranean island sunset, Van Gogh's *Sunflowers*, and Strauss's *Im Abendrot* are all beautiful is certainly something experienced by most of us as true, as demanded by the objects so described, and as authoritative and compelling as anything in science. They are also judgements on which there is surprising convergence among different people and at different times, in the making of which people certainly believe themselves to be speaking about qualities of the things not about feelings in their own breasts.

Indeed, that beauty, as we speak of it, is not just a projection, a gilding and staining by internal sentiment, is strongly suggested by the experience most of us have had at one time or another of learning that something is beautiful without at the time actually appreciating it, and then with, care and acquaintance, coming to find it beautiful (a phenomenon memorably described by Proust). In other words, what *is* beautiful actually directs, shapes, and explains our sense (experience) of its being beautiful. It is not as if the feelings in us towards that thing existed prior to or independently of our inner assimilation of standards of true beauty, standards which we take to have an existence and a logic independent of empirically given feelings people have. As far as the logic of beauty goes, it is perfectly conceivable that a Ruskin could demonstrate to a people that there was something amiss about its taste, that, for instance, the aesthetic of Disney cartoons is not so much popular as low, garish, mawkish. Certainly, when we read Ruskin on Turner, or Nietzsche on Wagner, or Leavis on Eliot, we feel that what we have are attempts to do more than simply emote. These critics are pointing to features of the work which we are urged to accept; they *demand* a response of a certain sort: if we disagree—as in at least two of the cases we should—what we have to do is to point to countervailing features which override the original judgement. Note, incidentally, how discourse of this sort reinforces Kant's position on the singularity of aesthetic judgement: that—to stay with our examples—religiosity like Eliot's, but in hands other than his, or a neuraesthetic sensibility like Wagner's, but without his colossal grasp of structure, might well have had the dire effects Leavis and Nietzsche claimed to find in the work as a whole. While there are principles or, better, reasons which support and must support aesthetic judgements, an aesthetic judgement will always depend on the precise and unique way a particular set of principles bears on the specific case.

Reflection on the way discussions about the beautiful actually proceed shows clearly that neurological accounts of the perception of beauty are necessarily inadequate. P. W. Atkins has written of the near universal recognition of the golden section as being 'like the resonance of an electronic circuit to a distant transmitter, when a certain frequency can induce an enhanced power'.[10] Particular circuits in the brain respond to the scanning of golden sections, the brain itself responds in 'an enhanced manner' and our response is 'the one that normally correlates with perceived "beauty"'. What is perceived as beautiful, on this 'scientific' account of beauty is what meshes with our brain circuitry, which is itself a product of our DNA, a function of our evolutionary history.

I do not want to deny that at some level perception of aesthetic properties is based in our physical constitution, our brain structure, and hence, indirectly, in our evolution. Thus, E. O. Wilson's claim that our predilection for Claude-type classical landscapes may have its origins in our ancestors' liking for water combined with the visual interest of cliffs and rocks on the savannahs they inhabited is suggestive (though, it is interesting to note, Ruskin found Claude's nostalgic

---

[10] P. W. Atkins, 'The Rose, the Lion and the Ultimate Oyster', *Modern Painters*, 2/4 (Winter 1989), 50–5. I replied to Atkins in 'Simplicity, Science and Art', in *Modern Painters*, 3/1 (Spring 1990), 57–9.

classicism insipid).[11] But talk about brain circuitry and evolutionary origins can at most provide us with the material substratum of aesthetic perception. For, contrary to what Atkins's account suggests, it is not the golden section *per se* which we find beautiful. Indeed, I doubt very much that the geometrical figure on its own would engage anyone's aesthetic interest for long, if at all. Nor, as Ruskin's response to Claude's landscapes suggests, is it any pond or lake in any green field which is aesthetically pleasing.

The perception of beauty is not an automatic response to a generalizable stimulus. The same golden section can be discerned in many different objects, such as a prehistoric palace in Mexico, a temple in Greece, a sketch by Le Corbusier, the Palazzo della Cancelleria in Rome, and, of course, a figure in a Euclidean textbook. Even if each of them struck the perceiver as *balanced*, that is not the same as harmonious, let alone beautiful. It is only the *right* sort of balance that is harmonious or beautiful, or balance *in a particular context* (i.e. aesthetic judgements are, as Kant urged, both normative and singular). Even if modernist architects have forgotten it, Alberti's adage that the beauty of ornament must be added to the harmony of proportion remains true. But what is the *right* combination of proportion and ornament? How natural must our silvan glade seem so as not to be insipid, or how humanized must a landscape be to prevent its being dreary or desolate?

We are partly back with Kant's insistence on the non-generalizable nature of the aesthetic judgement, that it focuses on the precise detail and make-up of a particular object, and also with its normativity. It is about how this particular thing ought to be, a judgement elucidated through giving reasons and how they apply in the particular case. In this sort of discussion, appeals to brain circuitry or evolutionary history will be strikingly irrelevant, because they tell us nothing about how or why this bit of ornament in a building enhances or destroys its overall harmony.

To put this point another way, our experience of beauty, either in a work of art or in a natural object, is not like certain nerves or neurones being stimulated. It is not simply some appeasement of the senses or of the brain: it is not like hunger or thirst being assuaged. The experience always involves reflection and judgement. It involves at the very least a perception of the beautiful object as an object of such and such a sort, and as beautiful in such and such respects. And, as a result of

---

[11] Wilson, *Biophilia*, 109–11.

intentionality of the aesthetic judgement, critics of our judgement can then point to features of the object we have missed or comparisons we should make, which might then lead us to revise or refine or reject our original judgement, and so to modify the experience. There is nothing analogous in sensory stimulation, whether this is pleasurable or painful. Even though selective attention can modify an experience of pure pleasure or of pain, what is at issue is an experience unmediated by thought or judgement, not responsive to reasoning in the way judgements, including aesthetic judgement, are. We can have aesthetic judgements closely tied to or based on sensations, as in the case of wine-tasting, but in the element of judgement involved the experience takes on rational and cognitive dimensions transcending the purely sensory. In an aesthetic experience, at the very least the pleasure in question is not just experienced; it is perceived as having a value, and, if Kant is right, as having a universal value. It is because it is perceived as having a value that the judgement is thereby open to criticism and refinement, and at the same time the person making the judgement and having the experience sees his life as, to that extent, improved.

The neural circuitry account of the experience of beauty overlooks the intentionality of the experience and the role that reason-giving and justification play in refining our sense of the beautiful, and also the way that far more than the formal properties of things we find beautiful contributes to our sense that they are beautiful. A related point can be made about the account of our sense of beauty given in terms of artists and others engaging in ever more refined performances in order to achieve success over their rivals, and, more particularly, success with women. There is doubtless something in such an account as far as the motivation of individual artists goes, and the account certainly sits well within the general orbit of evolutionary explanation. Like Dawkins's postulation of memes what this Darwinian account of aesthetic activity fails to touch on at all, however, is why the rest of us value particular performances, particular types of artistic expression, and so are prepared to reward their creators. What we are told is part of the reason why people might be driven to artistic endeavour beyond the norm, but not what it is in the endeavour itself which either artist or audience finds worth while. It is, even more than the neural circuitry thesis, an externalistic account touching on some of the dynamics of the artistic community but telling us nothing about the meaning or justification of judgements of beauty.

The objectivist view of aesthetic evaluation, which I am here defending, has the advantage over its subjectivist sociological and neurophysiological rivals of not convicting us all of error in our everyday aesthetic judgements. That is to say, when we call a sunset beautiful, we unreflectively take ourselves to be speaking of the sunset and its properties. We do not, as Hume, Kant, and their followers maintain, take ourselves to be speaking about nothing in the object, or to be merely gilding and staining it with projected sentiment, nor do we think we are simply referring to the workings of neural circuitry or the idiosyncratic whims of fashions.

Hume and Kant not only hold an 'error' theory of aesthetic value. In so doing, they also deprive themselves of the most obvious explanation of our shared sense that at least some aesthetic judgements are both non-arbitrary and that experienced, knowledgeable, and sensitive judges of works of art tend over time to agree in their judgements. We do indeed, as Hume insists, have a strong sense that the best works of art survive time and cross cultures. After all, none of us is a classical Greek or a medieval Florentine, and yet we continually look to their art as exemplary. And, even with small knowledge, most of us can appreciate the fineness of Japanese painting, say, or of Islamic architecture.

When an inquiry is both about something objective and on the right lines, there will tend to be a convergence on the part of disinterested inquirers. And where an inquiry is about real objects, participants in it will concentrate on picking out features of the objects under consideration. As we have already argued, this is a feature of critical discourse which, *pace* Hume, spends more time discussing the qualities of the objects viewed than it does in expressing the critic's feelings about them.

Aesthetic Realism

What are the metaphysical consequences of analysing aesthetic judgements in objective terms? And how would an aesthetic realism relate to our evolutionary past, and whatever glimmerings of an aesthetic sense we find there?

In her essay 'Against Dryness' Iris Murdoch considers the metaphysical predicament of post-Kantian man (a consideration extended in *The Sovereignty of the Good*). She says 'we no longer see man against a background of values, of realities which transcend him. We picture man as a brave naked will surrounded by an easily comprehended empirical world. For the hard idea of truth we have substituted a facile idea of sincerity'.[12] Aesthetic sincerity is the direction indicated by a Humean analysis: we look into ourselves to uncover the precise amount of gilding or staining we contribute at any time. But this, as we have seen, is a travesty of the critical process, or, indeed, of aesthetic inquiry generally.

One possible interpretation of the objectivity of aesthetic value would be to see it in terms of a background of value, such as Iris Murdoch postulates, of standards to which our judgements ought to conform. The hard idea of truth against which we compare our judgements of Homer, Beethoven, Turner, and the rest would be a metaphysical fabric of value, something in or behind the empirical universe, and which our own aesthetic creations and perceptions occasionally and fleetingly reveal. Again, according to Iris Murdoch,

> Good art, thought of as symbolic force rather than statement provides a stirring image of a pure, transcendent value, a steadily visible enduring higher good, and perhaps provides for many people, in an unreligious age without prayer or sacraments, their clearest *experience* of something grasped as separate and precious and beneficial and held quietly, and unpossessively in the attention.[13]

For a Platonism of this sort the magnetism of aesthetic value would stem not, as it does for Hume and Kant, from our perceptions of aesthetic value stemming from within us. It would be rather that what is within us is itself attracted by and pulled to something beyond us, the Beautiful, or the Real under the form of the Beautiful.

Plato

In *Phaedrus* (250 ff.), Plato sees the perception of beauty on earth as a reminder of the beauty we knew before we fell to earth. He is speaking primarily of the beauty of human beings we are physically attracted to. There is in Plato little sense of other types of beauty; he speaks mainly of encounters with godlike faces or bodily forms. In such an encounter, the lover shudders with awe and reverence, as at the sight of a god. Then a strange sweating and a fever seizes him. Warmth enters his soul by reason of the beauty coming in through his eyes. His soul grows wings and aches, as does a teething child. But when he is

---

[12] I. Murdoch, 'Against Dryness', *Encounter*, 16 (Jan. 1961), 16–20; *The Sovereignty of the Good* (Routledge, London, 1970).

[13] I. Murdoch, *The Fire and the Sun* (Oxford University Press, 1977), 76–7.

parted from the loved one he goes mad, and is ready for any shame to get back to the beloved. Plato then describes the struggle in the soul between the two horses, the orderly one and the wanton one. If the good horse wins, the lover will sublimate his carnal desires, and both lover and beloved will enter into the ordered rule of the philosophical life, able at the end to take the first steps back on to the celestial highway, no more to return to the dark pathways beneath the earth.

Physical beauty, then, awakens other, more spiritual longings. It can open the eye and the soul to a different type of beauty altogether. But what is striking in Plato are the continuities: from carnal attraction to spiritual vision; from highly interested, even crazy possessiveness to calm and steady admiration. Reading Darwin about beauty and then reading Kant, it is hard to see that they are talking about the same thing. But they are, and Plato fills in the gaps in a way which, while mythical, does correspond to our experience of beauty—or, rather to some of the many things which ambiguously and in some mutual tension go to make up our experiences of beauty. That is to say, aesthetic experience—the appreciation of, being attracted by beautiful things— covers a whole continuum from something pretty close to sexual attractiveness right through to a sense that in perceiving something as beautiful we are reaching to the very core of existence. But we must stress that this is a continuum, as Plato teaches. Sexual attractiveness can lead to something disinterested, and, as we have observed, does in the hands of Titian and Velázquez. Equally, the beyond to which aesthetic experience points is not an immaterial beyond, or if it is, it is not a beyond we can get at independently of the work or object we are contemplating. There may indeed be something transcendent about *The Flaying of Marsyas* or *The Death of Actaeon*—certainly both these paintings speak eloquently and ambiguously about a (possibly spiritual) transformation of the body in and through its suffering, but what it is cannot be conveyed except through the images and the physical-cum-psychological experiences they refer to. No more, indeed, has the Turnerian sublime (as exemplified in *Hannibal Crossing the Alps*, say) any meaning outside the dissolution of the everyday in the power of nature. This, of course, raises a problem when artists attempt to go straight for transcendence, a problem I have discussed elsewhere in connection with Rothko, whose best work remains no more than a reminder, a hint, a sketch, and never the thing itself.[14]

---

[14] In 'The Real or the real, Chardin or Rothko?', in M. McGhee (ed.), *Philosophy, Religion and the Spiritual Life* (Cambridge University Press, 1992), 47–58.

Plato, of course, is not himself immune from the incorporealizing of beauty. At least in *The Symposium* (210d), he speaks (or Diotima speaks) of the lover of beauty turning his eye from the 'slavish and illiberal' devotion to 'the individual loveliness' of single boys or men or institutions, and setting sail on the 'open sea beauty', towards beauty itself, the universal beauty. Diotima's doctrine is that this is the final revelation, detached it seems from the mortal taint of the particular. None the less, it is only through appreciation of bodily beauty that the initiate can begin on the road to enlightenment.

The Platonic accounts of the ascent from the particular beauty of a particular boy to the love of beauty itself are full of difficulty. It is never, for example, clear just how the boy's beauty participates in absolute beauty. Plato's fatal step, at least in *The Symposium*, though, is where he tells us to think of a particular beauty as being 'of little or no importance' (210b), compared to what all beautiful things have in common. Plato wants to secure Kantian disinterestedness or detachment at the price of uninterestedness in any particular and our looking way beyond it. Kant here is surely right. While love of beauty is or can be disinterested, and in a way should be, it is realized only in attention to particulars. Disinterestedness means we are not obsessively and graspingly fixated on one individual. It implies a calmness and a readiness to focus on different exemplars, and indeed a Chardin-like readiness to find beauty in the mundane and the ordinary. But the very calmness and attentiveness in which beauty is most fully experienced is precisely an attentiveness to and appreciation of the detail of the individual work, which remains potent when the initial almost physical excitement declines. We do not need to think of sexual beauty to come to an appreciation of this point: we can think of how a music lover may soon tire of the *1812* Overture and grow into a love of the more cerebral works of Bach. (In saying this, I do not want to do down Tchaikovsky's *œuvre* in general: that would be to succumb to the exaggerated Platonism I am criticizing: for all his occasional vulgarity and over-emotionalism, Tchaikovsky remains a great composer, even in his popular works, as Stravinsky for one always acknowledged.)

Leaving aside its metaphysics of abstraction, Platonism brings out three things in our experience of beauty: one, that it is based in physical attractiveness, and hence closely linked at times to sexuality, the area in which genetically based physical attraction is most strongly marked; two, that in judging something as beautiful, we feel that we are guided by standards apart from our momentary feelings; and, thirdly, that in experiencing beauty we feel ourselves to be in contact with a deeper reality than the everyday. Each of these three ideas is, I would contend, constitutive of most people's aesthetic experiences. What Plato's philosophy of beauty attempts to do in part is to assure us that what we feel deeply about beauty in art and in nature rests on reality, and that it is not mere projection. In arguing, as I am doing, that our experience of beauty is not mere projection, are we committed to something like the full Platonic account as its condition of possibility? Or is there a *via media* between Hume's emotivism and a Platonic Beauty?

All this is very close to Kant:[15] the beautiful thing pleases *immediately*, not as being an example of something general; it does so *disinterestedly*: in judging something beautiful our imagination is acting *freely* (that is according to rational criteria and not as a mere response to a stimulus, in other words); and in so judging we are judging *universally* (intending general acceptance for what we say which is not merely an expression of subjective taste).

Kant is very clear about what the underlying significance of the aesthetic judgement would be if it were genuinely objective, particularly in respect of nature: that in perceiving part of the natural world as beautiful we would be 'pleading eloquently on the side of the realism of the aesthetic finality of nature'—that is, that beneath our perception of beauty, nature is organized in our interest (flowers, etc., 'selected for our taste').[16] *If* nature has been designed for our taste, then presumably much else in it must be *for us*. Continuing the thought, artistic beauty, too, involves a humanization of our surroundings, which is so much whistling in the dark if, in some sense, our world is not humanized or humanizable. (It is no coincidence that the ugly in art, and the trivial, goes hand in hand with irreligion.)

But Kant, though understanding what aesthetic realism *would* imply, denies it. He rejects any such realism in part he says because nature can produce many pleasing forms merely mechanically. This, no doubt, is true, though it does not explain why we are attracted to such 'merely mechanical' forms. Here, indeed, evolution might help, by showing that our faculties are produced by, and hence amenable to, the very same processes which produce the mechanical forms of nature.

More fundamentally, though, Kant thinks that an ultimately theological or teleological aesthetic realism of the sort he rejects would

<hr>

[15] Cf. *The Critique of Judgement*, sect. 59.   [16] Ibid., sect. 58.

tend to deny the freedom of our judgement. That is to say nature itself (or its Creator) would be dictating to us what we find beautiful, and the source of the judgement would not be in us, due to the free play of our imagination.

Kant is surely right in insisting that aesthetic judgements are not simply responses to stimuli, and that recognition of standards and the exercise of critical rationality are involved, as we have argued. But that in itself does not settle the matter as to whether at a deeper level what we—freely and rationally and through historical trial and error—come to put forward as the standards of the beautiful have their ultimate source in us, as opposed to being based in reality more generally. What Kant says about the freedom of the aesthetic judgement is, of course, analogous to what he says about moral judgement. There, too, he insists that its autonomy means that it has its ultimate source in us. The thought is that in making a moral judgement, we are exercising our rationality, and if what we said was simply forced on us from without, by some facts apart from us, then we would not be judging freely. Kant's aim in both aesthetic and moral case is to draw a contrast with judgements which are determined from without, as he believes judgements about the physical world are.

Kant underestimates the extent to which our judgements about the physical world are determined from within, the extent to which they depend on human response; but more significantly, he fails to consider the possibility that a type of judgement could be both dependent on some special mode of thought in us (and hence 'free') and yet picking out something real (and hence 'objective').

Psychologically or in the life history of each individual, the initial wellspring of our evaluations, moral and aesthetic are, indeed, as Hume and Kant argue, certain feelings in us. This explains the magnetism of value. But once we collectively and individually begin to refine our ethical and aesthetic reactions to things and to discuss them with each other, we start to think of the things themselves as good or bad, cruel or kind, beautiful or ugly. More important, we begin to envisage the possibility that we ourselves, or many people, or even most people might be wrong on such matters.

At this point, what might initially have seemed like a gilding or staining of the world on our part, seems more like a response to a world which is gilded and stained, and which in various ways we *discover* to be gilded and stained. To call a musical theme noble is not just to speak about the feelings it evokes. It is to draw people's attention to its features in a particular way. In such circumstances, aesthetic feeling or reaction may well follow the cognitive analysis, and will in any case be justified in terms of the analysis. In Kantian terms, the perceiver is judging, discovering, and cognizing, as much as he is legislating.

When this point is reached, as from the point of view of the race it was reached perhaps three or more millennia ago, as David Wiggins has argued a 'system of anthropocentric properties has surely taken on a life of its own. Civilisation has begun.' [17] That is to say, human beings have ceased to be moved merely by animal instincts and idiosyncratic urges, and have begun to see their evaluations as corrigible and decidable on the basis of conversations directing attention to real features of the world and its objects. Can we imagine the aesthetic responses of a peahen being refined and developed in this sort of way? No, because her responses, being far from disinterested, exist only as issuing in action. Once stimulated, there is no question of a response being right or wrong, no logical gap between the response and its grounds. There is only a genetically-based response which fires or fails to fire.

As children, each of us is introduced into conversations about beauty, and to the sort of considerations relevant to such conversations. We are introduced to a realm which is already in existence, and which presents itself as objective, and which can claim to guide our individual responses. It is, in a sense, an anthropocentric realm, depending on human responses. Kant was right about that. But its being anthropocentric in that sense does not preclude its being also objective in a full sense: as reflecting and eliciting certain genuine features of the world. Even though aesthetic (and moral) judgements are our judgements, it does not follow that our sensibility is not caused by and ideally in tune with deep aspects of reality. It *may* be the case as the Platonist would aver that we have been made for the world, and the world for us. Anthropocentric as it is, based in us as it is, it may be the case that aesthetic experience reveals something of the world's inner essence, and that it heals the rift opened up by objectifying, decentring science between us and the world.

Such indeed was the claim of Nietzsche in *The Birth of Tragedy*. Our highest dignity, he said, lies in the meaning given to us by works of art; and it is only as an aesthetic phenomenon that existence and the world are eternally justified. This is no doubt in part because art acts as a 'stimulus to life', as something full of energy, psychic and physical, and, to that extent, a manifestation of control and detachment. But its effectiveness as a stimulus to life must equally in part depend on the

[17] D. Wiggins, 'A Sensible Subjectivism?', in his *Needs, Values and Truth* (Blackwell, Oxford, 1987), 135–214, at 196.

possibility of taking a positive attitude to life, a possibility that must be based in reality if art is to reveal it. There were, in Nietzsche's view, two ways in which aesthetics can justify the world: the Apollonian, whereby the suffering of the individual was overcome by delight in the momentary appearances of things, and, more profoundly the Dionysiac, mainly music, in which we seek delight not in phenomena, but behind them. Dionysiac art

wishes us to acknowledge that everything that comes into being must be prepared to face a sorrowful end ... yet not to be petrified with fear ... for a brief moment we really become the primal essence itself, and feel its unbounded lust for existence and delight in existence. Now we see the struggles, the torment, the destruction of phenomena as necessary ... for all our pity and terror, we are happy to be alive, not as individuals but as *the* single living thing, merged with its creative delight.[18]

As is well known, Nietzsche later came to regret certain aspects of *The Birth of Tragedy*, fearing that in its adulation of Wagner it moved into a narcotic, life-denying Christianity, in which consolation and redemption were to be achieved by a denial of life, a denial of the body, and a denial of the will. Even worse, in its search for an art of metaphysical consolation it sanctioned mediocrity, democracy, and 'modern ideas'. What must first be learned is the art of 'this-worldly consolation'.

It seems to me that any 'aesthetic justification' must in a strong sense be this-worldly, and to do with appearances; Apollonian, in other words. In saying this we must include music. It would be quite wrong to follow Schopenhauer and Nietzsche himself in seeing music as not involving appearances, and going in some direct way to the essence of things. Whatever is communicated in music is communicated in and through its sounds. While not directly representational or mimetic, it wipes away appearance no more than do the visual arts. Indeed a musical performance demands a far greater paraphernalia of thoroughly corporeal instruments, performers, etc., to produce its essentially fleeting appearances than do the visual arts. Music is nothing without its sounds and their appeal to our sense of hearing. Beauty is and must be physically attractive, and whatever is revealed through beautiful things is an endorsement of the worth of the physical or material world. Not even *Parsifal*—especially not *Parsifal*, the work most detested by Nietzsche for its promise of metaphysical consola-

---

[18] F. Nietzsche, *The Birth of Tragedy* (1871), sect. 17.

tion through folly and compassion—achieves its purpose through wiping away physical sound and the beauty of physical sound.

In any case, it is a perverse interpretation of Christianity which sees it as offering purely metaphysical consolation. To the horror of Jews and Muslims, Christianity is Greek enough to postulate the Incarnation, the embodiment of the divine. As the Greek Fathers used to say, God became man so that man could become God. This implies that there is something specific about human existence of great potential value, a value which would be wiped out if our specifically human embodiment were also to be wiped out. So Nietzsche's interpretation of Christianity as essentially other-worldly is at best a half-truth, however much certain variants of Christianity may have encouraged it. There is also another sense in which—against Nietzsche—aesthetic consolation must be Apollonian. For Nietzsche the Dionysian involves a bacchic wiping away of individuality, a subsidence into the *single* creative thing, and, according to some, a descent into a pre-conceptual, even fascistic original myth in which the morality of respect for individuals is suspended. But such an abrogation of individuality and civilized value can hardly stand as a justification of *our* world, of the human world, which is based on a sense of the will, responsibility, and intentionality of individuals.

At the same time, the Dionysian sense that aesthetic experience sometimes transports us, piercing the veil of appearance is very strong, as Iris Murdoch suggests. Art can seem revelatory, just as it does seem to answer to objective standards. It can seem to take us to the essence of reality, as if certain sensitivities in us, and based on our physical constitution and its predilections, beat in tune with reality. It is as if our conscious delight in and appreciation of things external to us, both natural and artistic, are reflecting a deep and pre-conscious harmony between us and the world from which we spring. If this feeling is not simply an illusion, simply a matter of our gilding and staining reality, it may say something about the nature of reality itself, as responsive to human desires.

Materialists like Atkins explain the harmony between our aesthetic sense and aspects of the physical world in terms of our brains reproducing and responding to certain basic physical patterns which are deeply embedded in physical reality. Nietzsche himself on occasion analysed the attraction of music in terms of the way it speeds up our animal functions, with no reference to soul or spirit which in Nietzsche's view would be a Wagnerian illusion. Our experiences of beauty, though, promise far more than this, far more than what might be achieved indifferently by a drug, or a peacock's tail, or a warm bath.

As I have urged against the neural circuitry account, so against Nietzsche's bodily resonance account of beauty, reflective enjoyment is a key part of our experience of beauty. More than this, there is much in human aesthetic experience which is not in any sense a response to something straightforwardly attractive or comforting or harmonious. This aspect of our appreciation of beauty once more takes us far beyond a peacock tail response, engaging as it does judgements and perceptions beyond the sensuous present. Unlike the peahen, in our perception of something as beautiful, owing to the co-presence of self-consciousness and intellectual reflection, we are unable altogether to avoid a sense of the wider implications of our finding something beautiful. This is obviously true in the case of a lover being unable to forget his beloved's character or her status as married to someone else, even as he is dazzled by her beauty. But it is also an aspect of the shock we feel at reading of Elagabalus' supposed purely aesthetic reaction to the sight of his victims' blood on green grass or of Mussolini's nephew's pleasure in the beautiful patterns (like blossoming roses) made by bombs dropped on native horsemen in the Abyssinian war. And it also accounts for the difficulty morally earnest people sometimes find in perceiving anything as beautiful, given the urgency of the moral demands made on them and the prevalence of suffering, human and animal.

In our aesthetic experiences, to a greater or a smaller extent, our rational and critical faculties are engaged, but they are engaged in making us feel at home in a world that at other times they tell us is unknowable and alien. Nietzsche in his early work is right to describe at least one type of aesthetic experience as putting us in tune with the eternal delight of existence, and promising a moral and mental consolation, not just a physical cushion tailored to our senses. Aesthetic experience uplifts our desires and hopes as well as our bodies and it engages our intelligence as well as our feelings and sensations. Nietzsche is also right to concentrate on tragedy. If, as he believed, the world is justified as an aesthetic phenomenon, the redemption of individual suffering is a test case. What tragedy tells us—whether it is *Oedipus* or *Hamlet* or Brahms's Fourth Symphony or Titian's *Flaying of Marsyas*—is that though destruction is part of life and essential for creative rebirth, the essential individual, which each of us is, is neither wiped out nor harmed. Indeed, what is creative rebirth for, but for individual experience, a type of experience we know only as embodied?

Tragic art is possible and it may not require a context of dogmatic religion. But to show suffering as beautiful or as ultimately redeemable is to show the world as not ultimately alien, and ourselves as not necessarily alienated. It would also show that our constitution, physical and rational, sensual and reflective, is of some value in revealing the essence of the world. But, how could we think of an aesthetic justification of experience, that really was a justification and not just a momentary narcotic, unless our aesthetic experience was sustained by a divine will revealed in the universe, and particularly in our experience of it as beautiful? It is precisely at this point that many or even most will draw back. Aesthetic experience *seems* to produce the harmony between us and the world that would have to point to a religious resolution were it not to be an illusion. But such a resolution is intellectually unsustainable, so aesthetic experience, however powerful, remains subjective and, in its full articulation, illusory.

This is a dilemma I cannot solve or tackle head on. I will, instead, end this chapter, extravagant enough as it is already, with two further thoughts. The first is that if the line of thought of the last few paragraphs has anything in it at all, then Orthodox icons, however beautiful, and however religious, are only partial revelations of beauty in that they articulate only a part of human life and our experience of beauty. If there is any sense in which the world is justified as an aesthetic phenomenon, then, as suggested in Plato, it will not just be our specifically religious moments. Asceticism denies too much of our life and our sense of beauty, which is directed not only to the immaterial. Aesthetics, by contrast, even if it points to a transcendence of the everyday, has a transcendent *humanism* (and perhaps has more affinity with the medieval Islamic view of heaven than with the Augustinian hatred of the flesh).

The second thought grows out of the first, but has a Kantian resonance. Our knowledge-seeking faculties suggest to us that the world contains much that is beyond our understanding. Certainly the world exists apart from us and independently of us. Much in it is mysterious and problematic to us. Also, the very enterprise of science teaches us to think of the world as going on without us, and we ourselves as no more than a trivial contingency within it. Humanly speaking, the scientific world, or its image, is meaningless at the same time as the scientific would relegate the humanly meaningful to secondary status. Morality, too, leaves us with profound and insoluble problems, particularly in a world full of shortage and suffering, and in which our duties (if such they be) to those not close to us cannot effectively be discharged, at least not while we as individuals lead reasonably human lives. Kant was, of course, led to seek extra-mundane solutions to the problems thrown up by our moral sense.

Where, though do we get a sense that, despite the problems of alienation thrown up by science and morality, we are nevertheless at home in the world? Where else, except from a sense, at times strong, at times fleeting, that despite all the horrors and dilemmas and problems, the world and human life are beautiful, and that this sense is not mere projection on our part? From my point of view it is above all in aesthetic experience that we gain the fullest and most vividly lived sense that though we are creatures of Darwinian origin, our nature transcends our origin in tantalizing ways.

8

Conclusion

When we look back over the ground we have covered in this book, it will become apparent that in human activity there are myriad ways in which we are led to transcend our evolutionary origins. We have focused particularly on what I have, on a number of occasions, referred to as the normativity of the mental: that is to say that in pursuing knowledge, morality, and the beautiful we are led to seek goals which have nothing to do directly with survival, and which may at times militate against survival. To this extent the Platonic reflections of our introductory chapter are vindicated.

In talking of the normativity of the mental, though, I do not want to embrace a full-blooded Platonism. If, for purposes of convenience, we speak of truth, goodness, and beauty—roughly corresponding to the tripartite division of the core of the book—it remains true that our pursuit of these goals is formed and, to an extent, constrained by our embodiment and our social background. Thus, I have been at pains to emphasize the extent to which our pursuit of knowledge is based in our sensory faculties and, in Chapter 5, I have defended the validity of secondary quality perception against the supremacy of the absolute view of science. It is also worth remarking on the contingency of our actual science, based as it is in its history of actual problems, insights, and interests.

In Chapter 6, I argued similarly that our moral sentiments are founded in the primitive feelings Darwin describes in *The Descent of Man*, and our actual beliefs and practices are rooted in the traditions handed down to us. On beauty in Chapter 7, I stressed its basis in our embodiment, particularly, following Darwin, in our sexuality, and also in our secondary quality perception.

Against any Platonism which seeks to cut loose completely from our embodiment or our practices, I would maintain that our normativity, our aim of reaching universally acceptable rational goals, always operates against and within what we are given, in nature and in history.

But a normativity, which seeks to validate what we are given, and, where necessary, to transcend it does operate.

Thus in our knowledge-seeking we do seek something more than beliefs conducive to survival and reproduction. We seek truth for its own sake. This quest for truth in respect of the causal mechanisms underlying the empirical world takes us into the abstractions of modern science. These certainly transcend the sensorily given and many of these have little bearing on survival and reproduction. And if, in Chapter 5, I have sought to vindicate the validity of the sensorily given, this (in so far as it has been successful) has been based on an argument whose cogency should be acceptable even to rational beings with a different sensory make-up. I have not simply asserted the validity of our sensory make-up, but have sought rather to develop a standpoint from which the status of different sensory fixes on reality might be recognized as genuine insights into reality, even though they are not fundamental from a causal point of view.

Then again, in considering morality and politics I have emphasized the manner in which our rationality, our quest for reasons to justify what we do, together with the phenomenon of the gaze, conspire to shake us out of any dogmatism regarding our values. This may, indeed, leave us with an unresolved tension between our lives as they are lived and liveable, and what might seem to be the demands of a universal morality. There is also the tension, alluded to at the end of the last chapter, of aesthetic experience: that it hints at a transcendent hope our reason would deny.

At least as things stand, I have no means of resolving the tension, given that we are implicated both in the way things stand now and in how it seems they ought to be. All I can say in mitigation at this point is that to resolve the tensions by fiat would be both arbitrary and unfortunate. That is to say, simply to acquiesce in the way things are would be a denial of our rationality, and an admission that in the end all we are left with in the realm of human conduct is some kind of Thrasymachean (or is it Rortyan?) combination of rhetoric and/or persuasion. The moral, social, and political cost of such acquiescence—even if in practice it turned out to be acquiescence in the sort of left-liberal-feminist consensus which Rorty for one seems to favour—is surely unacceptable, if only because it amounts to an admission that thought and reason have no central role to play in areas where, in the end, they are our only hope, our only recourse. On the other hand, to attempt simply to sweep away custom and tradition in the name of rationality and without any clear idea of how to replace the felt and lived allegiances to systems of value embodied in the customary and the traditional, is itself a form of obscurantism. We need to remember that custom and tradition do embody a deposit of wisdom and learning from experience which we should not disregard. In general, in this area it seems better to live with the tension, in the hope that its working out will be creative, rather than to plump dogmatically for one side.

More doubts will be raised about the universality and transcendence-intimating nature of beauty even than about morality and social policy. Despite everything that has been said in the penultimate chapter on the topic, the suspicion will remain that beauty is essentially particular, and its recognition in specific instances confined to specific times and cultures. On the latter point what we are faced with here very directly is the dialectic which arose at the end of the eighteenth century, that between defenders of universal standards of reason and taste and those, like Herder and Fichte, who emphasized the rootedness of taste and style, and notions of beauty, in particular cultures and traditions.

Against this relativism, though, if our reflections on the normativity of human thought and evaluation amount to anything at all, so central a feature of human life as aesthetic appreciation could not be locally insulated from standards with wider application. We have already considered the Kantian view on the judgement of taste, as involving claims to universal attention and agreement. In practice, though, to those of us confronted at the end of the twentieth century with a plethora of artistic styles, national traditions, and standards of taste, the Kantian view can seem wilfully out of touch and unreal. It is just not the case that standards in this area are or could be transcultural: the history of art and taste, it will be said, shows the opposite, and in so showing, undermines the Kantian argument (and also, incidentally, itself—except as a purely anthropological or archival exercise, laying out differences and developments for non-evaluative classification).

But does it? It is worth reflecting here on the biography of one who lived through an enthusiasm for the particular, to reveal himself eventually as one of the great exponents of transcultural aesthetic understanding and practice, namely Goethe. In considering Goethe, we may of course reflect on what any of us is doing, to a lesser degree no doubt than Goethe, when we visit an exhibition of Islamic textiles, or Japanese woodcuts, or Byzantine icons, or Benin bronzes, and appreciate what we see.

In 1772 Goethe wrote an essay on German architecture. This was shortly after he had met Herder, and had been impressed by his enthusiasm for the primitive and also by his arguments that each culture has its own meaning and value, irreducible to and unintelligible in terms of the meanings and values of other cultures. Goethe's essay is shot through with Herderian resonance, and in it he praises the Gothic architecture at the expense of the classical, which was paradigmatically the *international* style of eighteenth-century Europe. In the form of an address to Magister Ervinus (Erwin von Steinbach), the architect of Strasbourg Cathedral, Goethe comments how the Italian would describe the Minster as in niggling taste, with the Frenchman childishly babbling 'Puerilities' while triumphantly snapping open his snuffbox, 'à la Greque'.[1] Moreover,

the first time I went to the Minster my head was full of the common notions of good taste. From hearsay I respected the harmony of mass, the purity of forms and I was the sworn enemy of the confused caprices of Gothic ornament . . . no less foolish than the people who call the whole of the foreign world barbaric, for me everything was Gothic that did not fit my system . . .

And yet, on actually seeing it, 'how surprised I was when I was confronted by it', and, as a result of the magical and above all natural impression given by all its ornament and detail, he quickly came 'to thank God that he can proclaim that this is German architecture, our architecture. For the Italian has none he can call his own, still less the Frenchman.' The German Gothic is a characteristic art. As such it is the only true art 'unadorned by, indeed unaware of, all foreign elements [and] whether it be born of savagery or of a cultivated sensibility, it is a living whole'. Hence, among different nations you will see countless different degrees of characteristic art, and in the case of Strasbourg, what we have is an example of the 'deepest feeling for truth and beauty of proportion, brought about by the strong rugged German soul on the narrow, gloomy priestridden stage of the *medii aevi*'. Even as late as 1823, Goethe endorsed the favourable impression the Minster had made on him in 1772,[2] but by then his attitude to national culture had been somewhat transformed, as we can see from a conversation Goethe had with Eckermann on 21 January 1827.[3]

---

[1] In 'On German Architecture (1772)', in *Goethe on Art*, ed. and trans. John Gage (Scolar Press, London, 1980), 103–12. Cf. also Alain Finkielkraut, *The Undoing of Thought* (Claridge Press, London, 1988), 96 ff., on which my account of Goethe draws.

[2] Cf. Goethe's 'On German Architecture', in Gage, *Goethe on Art*, 118–23.

[3] J. P. Eckermann, *Conversations with Goethe* (Everyman's Library, London, 1970).

Goethe had been reading a Chinese novel, expecting to be struck by its strangeness and difference, but he had been struck instead by its closeness in theme and treatment to his own *Hermann and Dorothea* of 1797, and also to the English novels of Richardson. Of course men were rooted in particular places, and up to a point creatures of their traditions, histories, and geographies, but these divisions and fragmentations could be transcended, particularly through art:

I am more convinced that poetry is the universal possession of mankind, revealing itself everywhere and at all times in hundreds and hundreds of men. One makes it a little better than another, and swims on the surface a little longer than another—that is all . . . we Germans are very likely [not] to look beyond the narrow circle that surrounds us. I therefore like to look about me in foreign nations . . . The term 'national literature' does not really mean much today. We are moving towards an era of universal literature, and everyone should do his best to hasten its development . . . While we value what is foreign, we must not bind ourselves in some particular thing, and regard it as a model.

And he says that if we want a model we must look to the ancient Greeks, where the beauty of mankind is universally and constantly represented. The rest we must look at historically, taking for ourselves what is good, as far as it goes. In March 1832, shortly before his death, Goethe appeals even more forcefully to universal standards: 'as a man, as a citizen, the poet is bound to love his native land . . . but the native land of his poetic powers and poetic action is the Good, the Noble, the Beautiful, which have no particular province or country, and which the poet seizes on and forms wherever he finds it.' The poet, for Goethe, is like an eagle, hovering and gazing over whole countries, it being of no consequence to him whether the hare he pounces on is running in Prussia or Saxony. What then could be meant by love of one's country, other than setting aside the narrow views of his countrymen and so enobling their feelings and thoughts?

The Herderian position on culture will deny this, and insist that each individual becomes even more rooted in his or her own particular culture, though whether this is intended as an epistemological or a moral-cum-political claim, or a mixture of the two, is not always entirely clear. As an epistemological claim, it is, of course, refuted by the experience of Goethe with the Chinese novel, by the experience of anyone today who reads Homer with some understanding, and at a significant but lower level, by the universal popularity of originally American pop music. It is this last phenomenon, indeed, that might lead many to see something attractive in Herder's position from a moral or aesthetic point of view. Doesn't universalism lead to a destruction of much that is worthwhile, and to an abrogation of value judgements, leading in the end to a universal mediocrity? Here, I think, one has to beware of a sentimental form of primitivism. Vulgarity and worse do not occur only in Disneyland but may be evident in village culture. And, as the example of Goethe himself shows, transcultural borrowing and assimilation can be conducted on the highest level.

In fact, on analysis, it turns out, perhaps at first sight paradoxically, that a Goethean position is far more conducive to the making of value judgements and to the preservation of the worthwhile than Herder's. For it was Herder who denied the possibility of making valid judgements across cultures and who thus in effect deprived the admirer of a peasant culture of any firm ground on which to criticize the incursions of satellite television or Coca-Cola. In a Herderian dispensation, he will appear simply to be defending the picturesque against the popular, having been deprived of any notion of any objective value on which to argue his case.

Goethe, by contrast, does appeal to a universal sense of value, which is manifested in but transcends particular cultures. He is neither committed to the defence of everything in a culture he admires overall, nor need his defence of ancient cultures against modernization be simply a lament for the passing of the old: he can point to universal values its passing will offend. His complaints against the effects of industrialization need not be sheerly Luddite, mere obstinacy in the face of universal progress. Paradoxically, then, it is Herder, with all his affection for the particularity of primitive cultures, who reduces their defender to the position of the Luddites castigated by modern technophiles. Goethe, with his references to the Good, the Noble, and the Beautiful, and with his extolling of the man who stands above nations (Eckermann, 14 March 1830), opens up the logical space needed to rank and assess cultures and their elements. There is, then, no contradiction between an appeal to universal values and a defence of particular aspects of particular cultures. Quite the contrary, what would be contradictory would be to attempt such a defence while denying any transcultural standards of assessment.

If such a denial is what is meant by pluralism, then F. R. Leavis's strictures on pluralism are well taken: ' "Pluralism" denotes a sitting-easy to questions of responsibility, intellectual standard and even superficial consistency, the aplomb, or suppleness being conditioned

by a coterie-confident sense of one's own unquestioned sufficiency—
or superiority.'[4] Leavis's target here is Noel Annan: whether his sally
is justified in that case, it could certainly be said that the sort of non-
judgemental pluralism he criticizes received an early exposition in
Herder, and after various incarnations on the political right and on the
political left, is with us today as those forms of multiculturalism which
would refuse to allow one to evaluate the worth of other cultures or
their customs. In most, if not all of its incarnations, there has also been
the sense from proponents of pluralism that they consider themselves
somehow superior, intellectually and morally, to those benighted folk
within particular cultures, who insist on the unique correctness of
their views.

But one does not have to subscribe to the Enlightenment view of
human progress or deny the fact of our own situatedness in particular
cultures to think of values as universal. Indeed, a proper account of
judgement would be that it is both individual in expression and formulation, and, in intention, universal; while judgement may indeed
depend on backgrounds of communal agreement, it cannot in the final
analysis be communal. Any judgement is made by an individual who
then takes some responsibility for it. As Leavis himself puts it, speaking of judgement about poems, 'a judgement is personal or it is nothing, you cannot take over someone else's'. And in describing the
process by which critical judgements are then discussed, amended, and
ratified, he goes on to speak of 'the collaborative-creative process in
which the poem comes to be established as something "out there", of
common access in what is in some sense a public world'.[5] Leavis goes
some way to capturing both the individual and the universal poles of
judgements of all sorts—that they are made by individuals, while seeking general or universal agreement. He then goes on to make the
important point that a culture can exist only in so far as it provokes
renewed responses from the individuals in it, 'who collaboratively
renew and perpetuate what they participate in—a cultural community
or consciousness'. In other words, a culture, if alive, is always changing in response to the judgements and reasons of those who live in it.

What all this amounts to is that while individuals stem from their
cultures, are rooted in them, if you like, they are never completely
determined by them. They are individuals, and in making their judgements, they appeal implicitly at least, to standards of correctness
which stand above the particular, and by which the particular is

[4] F. R. Leavis, *Nor Shall My Sword* (Chatto and Windus, London, 1972), 32.
[5] Ibid. 62.

judged, whether the particular is the individual person or his or her society. And there are some types of judgement no human being can avoid having to make. These judgements include particularly those about central values, moral, aesthetic, and even epistemological, in so far as epistemology bears on ultimate metaphysical and religious commitment.

While reflection and personal judgement are key aspects of any human life, the precise significance they have in particular cases will vary according to the nature of the individual and according to the type of society to which he or she belongs. Without wishing to minimize the personal responsibility an individual has for his or her beliefs and values—for this is a purely logical requirement arising simply from acquiescence as a rational being in given beliefs or values—we must recognize that the working out of this logic and this responsibility can and does take many different forms. At one extreme there is the individual who appears never to question anything, and simply accepts what he is brought up to hold. At the other, there is the Hamlet-like figure—or, more bathetically, in our own day the Woody Allen type—in whom reflectiveness and the desire to think correctly has extinguished all natural moral reactions and has all but paralysed the power of action.

As Oakeshott, above all, has argued, in our contemporary moral consciousness, reflectiveness has become predominant, and intellectuals particularly have failed to recognize the importance of grounding reflectiveness in habits of moral behaviour (and doubtless similar points could be made about the attempt to intellectualize scientific method). From the Oakeshottian perspective, as with Aristotle, Hegel, and Bradley, reflection is always to be seen as reflection on existing practice. Ideals which emerge in the course of reflection are not to be seen as floating completely free of the practices in which they have their role and significance. There are to be no ideals which do not have their intimations in existing practice, no heavenly city to which we can aspire which is not already part of our earthly city. And we must always be wary of taking apparently heavenly ideals and applying them out of context: the upshot, as seen so often in the twentieth century, is likely to be hell on earth, rather than a new heaven.

More generally, our inquiries and reflection, whether on values or on factual matters share in the conditionality of our practices if only because they are actually part of these practices. On the other hand, and in so far as rational reflection *is* part of our practices, there is within our practices the wherewithal to transcend total particularity, and to move into a more universally acceptable context. In response to critics who find his appeal to traditional wisdom over-optimistic in a world with a chaos of different norms, Gadamer wrote that

the displacement of human reality never goes so far that no forms of solidarity exist any longer. Plato saw this very well: there is no city so corrupted that it does not realise something of the true city: that is what, in my opinion, is the basis for the possibility of practical philosophy.[6]

Gadamer is, it seems to me, both too optimistic and too pessimistic. If Vaclav Havel is to be believed, communism nearly succeeded in extinguishing most meaningful forms of actual human solidarity in pre-1989 Prague. On the other hand, even if a particularity is through and through corrupt, in human life and discourse, and in practice more generally, the possibility of what Gadamer calls practical philosophy remains, if only because of our in-built rationality. What Gadamer makes appear a fortunate and possibly uncertain outcome of human history is in fact part of the very structure of our mental life and of any practices in which that life is able to develop (which does not, of course, mean that there cannot be political and social circumstances highly unconducive to such life).

We come here to the core meaning of human freedom, individual and social, and to the multifarious ways in which our lives and histories are unconstrained by any givens, Darwinian or otherwise. The underlying reason is that in human language and thought we are always assessing our actual beliefs and values against the way things should be. As self-conscious agents, we realize that what we actually think and believe may not be how other agents think and believe or, more to the point, may not be how things actually are or should be.

In this regard an unthinking acquiescence in how things are will always amount to a repudiation of human freedom and rationality. To be told by Burke that individuals would do better to avail themselves of the general bank and capital of nations and ages than to trade each on his own private stock of reason is, as we have seen, good advice. Similarly, we can take on board the lessons of sociobiology and sociology to the effect that all kinds of beliefs, customs, and practices may perhaps surprisingly have survival value; and this may indeed tell us something about their relation to the truth.

A shallow rationalism might ignore considerations of this sort, and pit a private stock of reason against the wisdoms embodied in instinct,

---

[6] From a letter Gadamer wrote to Richard Bernstein, and quoted in the latter's *Beyond Objectivism and Relativism* (University of Pennsylvania Press, Philadelphia, 1985), 263–4.

experience, and tradition: but there is nothing *per se* anti-rational or irrational in seeking to elucidate wisdom of this sort. To the contrary, the impetuosity of present desire masquerading as reason can be as much a denial of human freedom and imagination as a Rusean or even a Burkean critique of certain thoughts just because our genes or our society allegedly render them impossible or undeliverable.

If there is anything true in the present work, it would be that the relationship between genes and sociology, on the one hand, and our thoughts and ideals, on the other, is not like that. In having and expressing thoughts and ideals we are implicitly invoking the possibility and even at times the necessity of putting into question the deliverances of our genes or of our society. To attempt to read the true or the valuable or the beautiful straight off from evolutionary facts or historically given ways of going on is to excogitate a hopelessly hybrid monster. What should be subservient to norms of rationality is here being compressed into the unnormativity of pure fact.

For, as human beings, we are not the unwitting victims of purely natural or historical processes. Though we are indeed the products of natural and historical processes, our lives are largely lived on the level of what Oakeshott calls *intelligibilia*: that is to say practices through which we endow our behaviour with meaning. Built into the notion of a practice is reason-giving, and, as we have argued extensively, reason-giving takes us into a realm unconstrained by what is, reaching out to what should be.

Language and excogitated expressed beliefs, like our practices, and as crucial aspects of many of our practices, certainly conduce to survival. They certainly have had and continue to have an instrumental role, and can be seen in Darwinian terms. But, given that both have normativity and rationality built into them, they—like many of our practices—are not purely instrumental. They open up goals and aims and standards of their own, which then regulate and liberate our activity, in ways which are quite un-Darwinian.

As already suggested in connection with Leavis, the given—be it biological or sociohistorical or, more probably, an amalgam of both—is a condition both of individual will and of any meaningful activity. But in so far as individual and social activity involves appeal to norms of one sort and another, the given does not constrain our activity; rather, what is given contains within itself the means to self-transcendence. To speak once more in Oakeshottian terms, any human practice, to the extent that it is not merely mechanical and repetitive involves appeal to norms and standards, whose initial meaning is learned in learning the practice. But, then, after practice in the practice, what Oakeshott has called a 'prevailing educated sensibility' will grow up, which will allow for the development of the practice and encounters with other related practices, even from other cultures and ages. This development will not preclude but will at times include discussion of the rightness and applicability of the very norms and standards constitutive of the very practice within which the discussion initially arises.

In a sense, we can then say that, while in many ways our activities and practices are contingent and conditional, the modes of acting, thinking, and judging which are intrinsic to these activities and practices provide us with the means of transcending our present contingency and conditionality—though, of course, any significant transcendence of this sort depends on our having grasped the meanings and norms contained within the practice in question. What is being advocated here is not a post-modernistic insouciance towards any particular traditions or practice: the type of universality and transcendence we might hope for is one which demands in Goethe-like fashion a sustained immersion in the practices being transcended, a serious insight into the horizon being fused.

In a way, the post-modernist shares with the Rationalist the thought that human activity can be rootless: perhaps that it is best if it is rootless. While the general thrust of this book is by no means anti-religious, and while it is implied in Chapter 1 that its thesis is hostile to any form of materialistic naturalism which would see us purely in terms of physics or biology or any of the special sciences, its perspective is naturalistic and historical to the extent that it can conceive of no form of human life completely severed from our material and historical make-up. Our knowledge-gathering is through our perceptual make-up and cannot cut entirely loose from it, to the extent that I cannot conceive of a significant downgrading of secondary qualities. Our moral and social thinking is rooted in facts about our feelings, in the phenomenology of the gaze, and in the possibilities for effective reasoning these open up. Our aesthetic sense is tied to our sensibility, even while intimating transcendence of the limits it imposes.

We may speculate about just what are the conditions which allow the degree of normativity to enter the physical and biological world which we find in human life and just what the implications of this are for us as individuals and as a species. After all, as Darwinians like John Maynard Smith freely admit, there is on Darwinian principles simply no source whence absolute morality can come. Dawkins himself continually emphasizes that if, as he does, one wants to build a society in which 'individuals co-operate generously and unselfishly towards a common good, you can expect little help from biological nature'.[7] Similarly regarding truth, Rorty has argued that the notion that we are orientated not just towards our prosperity, but towards truth is as 'un-Darwinian' as the idea that every human being has a built-in moral compass. All this is correct; from a Darwinian perspective, truth, goodness, and beauty and our care for them are very hard to explain.

But they exist, at least in the sense that they condition and direct much of our activity. They are also implicit in our possession of rational and social self-consciousness in forms which are not reducible to Darwinian analyses and are not explicable in terms of Darwinian drives.

For some, speculation about the origin of our non-Darwinian concerns would take a religious direction, and they would doubtless take heart from the suggestion canvassed at the end of the last chapter as to the nature of aesthetic experience, at once particular and embedded, and reconciling that aspect of ourselves to what might be thought of as our striving for some transcendent guarantee and consolation. But, whatever direction such speculation takes it cannot be allowed to play down the significance of our embodiment or our history. It should not conceive of us operating outside a particular physical and historical situation, nor that such a thing would, were it possible, be desirable. For in cutting loose from our history and our embodiment, the judgements we make would lose the very significance and content our situatedness alone makes possible.

On the other hand, one moral to be drawn from this study is that Darwinism, if applied to our forms of intellectual, moral, and aesthetic life, is indeed a dangerous idea, as Dennett at least recognizes. For even though we and our capacities may have evolved in Darwinian ways, once evolved we and our capacities take off in quite un-Darwinian ways. It is not just that Darwinian analyses strike at the basis of our sense of self and at our self-respect, though they do that. It is rather that the account that they give of ourselves and our capacities involves a radical and unsustainable redescription of what we are and what we do.

---

[7] These words, from the very beginning of *The Selfish Gene* (Paladin, London, 1978), 3, are characteristic of Dawkins's very consistent views on morality and evolution. He ends *The Selfish Gene* by stating that 'pure, disinterested altruism' has never before humanity existed in the whole history of the world (ibid. 215). I would concur and would conclude that this simply emphasizes the *irrelevance* of Darwinism to morality.

Selected Bibliography

ARISTOTLE, *Nicomachean Ethics.*
—— *Politics.*
BURKE, E., *Reflections on the Revolution in France* (Penguin, Harmondsworth, 1986).
CRONIN, H., *The Ant and the Peacock* (Cambridge University Press, 1993).
DARWIN, C., *The Descent of Man* (John Murray, London, 1898).
—— *The Origin of Species* (Penguin, Harmondsworth, 1984).
DAWKINS, R., *The Blind Watchmaker* (Longman, Harlow, 1986).
—— *The Selfish Gene* (Paladin, London, 1978).
DENNETT, D., *Darwin's Dangerous Idea* (Allen Lane, London, 1995).
DEWEY, J., *Experience and Nature* (Dover Publications, New York, 1958).
—— *Human Nature and Conduct* (Modern Library, New York, 1930).
EDELMAN, G., *Bright Air, Brilliant Fire* (Penguin, Harmondsworth, 1994).
FINKIELKRAUT, A., *The Undoing of Thought* (Claridge Press, London, 1988).
GOLDMAN, A., *Epistemology and Cognition* (Harvard University Press, 1986).
GOULD, S. J., *Ever Since Darwin* (W. W. Norton, New York, 1977).
—— *The Panda's Thumb* (W. W. Norton, New York, 1980).
HAYEK, F., *The Fatal Conceit* (Routledge, London, 1988).
HEGEL, G. W. F., *The Phenomenology of Spirit*, trans. A. V. Miller (Oxford University Press, 1977).
HUME, D., 'Of the Standard of Taste', in *Hume's Ethical Writings*, ed. A. MacIntyre (Collier Books, New York, 1965).
KANT, I., *Critique of Judgement* (Hafner, New York, 1966).
LORENZ, K., *Behind the Mirror* (Methuen, London, 1977).
MILLIKAN, R., *Language, Thought and Other Biological Categories* (MIT Press, Cambridge, Mass., 1984).
MUNZ, P., *Philosophical Darwinism* (Routledge, London, 1993).
NAGEL, T., *The View from Nowhere* (Oxford University Press, 1986).
OAKESHOTT, M., *Rationalism in Politics* (Methuen, London, 1962).
PAPINEAU, D., *Philosophical Naturalism* (Blackwell, Oxford, 1993).
PEIRCE, C. S., *Selected Writings*, ed. P. Wiener (Dover Publications, New York, 1958).
PLATO, *Phaedo.*

PLATO, *Phaedrus*.

POPPER, K. R., *Objective Knowledge* (Clarendon Press, Oxford, 1972).

QUINE, W. V., *Ontological Relativity* (Columbia University Press, 1969).

RADNITZKY, G., 'An Economic Theory of the Rise of Civilization and its Policy Implications', *Ordo*, 38 (1987), 47–89.

—— and BARTLEY, W. W. (eds.), *Evolutionary Epistemology: Theory of Rationality and the Sociology of Knowledge* (Open Court, La Salle, Ill., 1987).

RESCHER, N. A., *A Useful Inheritance: Evolutionary Aspects of the Theory of Knowledge* (Rowman and Littlefield, Savage, Md., 1990).

RIDLEY, MATT, *The Red Queen: Sex and the Evolution of Human Nature* (Penguin Books, Harmondsworth, 1994).

RUSE, M., *Taking Darwin Seriously: A Naturalistic Approach to Philosophy* (Blackwell, Oxford, 1986).

—— and WILSON, E. O., 'Moral Philosophy as Applied Science', *Philosophy*, 61 (1986), 173–92.

SARTRE, J.-P., *Being and Nothingness*, trans. Hazel Barnes (Methuen, London, 1969).

SCRUTON, R., *An Intelligent Person's Guide to Philosophy* (Duckworth, London, 1996).

SMITHURST, M., 'Popper and the Scepticisms of Evolutionary Epistemology, or, What Were Human Beings Made For?', in *Karl Popper: Philosophy and Problems*, ed. A. O'Hear (Cambridge University Press, 1995), 207–24.

SOBER, E., *The Nature of Selection* (MIT Press, Cambridge, Mass., 1985).

STOVE, D., 'So You Think You are a Darwinian?', *Philosophy*, 69 (1994), 267–78.

TRIGG, R., *Ideas of Human Nature* (Blackwell, Oxford, 1988).

DE WAAL, F., *Good Natured* (Harvard University Press, 1996).

WARNOCK, M., *Imagination and Time* (Blackwell, Oxford, 1994).

WIGGINS, D., *Needs, Values and Truth* (Blackwell, Oxford, 1987).

WILLIAMS, B., *Ethics and the Limits of Philosophy* (Fontana, London, 1989).

WILSON, E. O., *Biophilia* (Harvard University Press, 1984).

WITTGENSTEIN, L., *Philosophical Investigations* (Blackwell, Oxford, 1953).

Index